Memoirs of the American Mathematical Society

Number 261

Janina Kotus, Michał Krych
and Zbigniew Nitecki

Global structural stability
of flows on open surfaces

Published by the
AMERICAN MATHEMATICAL SOCIETY

Providence, Rhode Island, USA

May 1982 · Volume 37 · Number 261 (third of 4 numbers)

MEMOIRS of the American Mathematical Society

This journal is designed particularly for long research papers (and groups of cognate papers) in pure and applied mathematics. It includes, in general, longer papers than those in the TRANSACTIONS.

Mathematical papers intended for publication in the Memoirs should be addressed to one of the editors. Subjects, and the editors associated with them, follow:

Real analysis (excluding harmonic analysis) and applied mathematics to PAUL H. RABINOWITZ, Department of Mathematics, University of Wisconsin, Madison, WI 53706.

Harmonic and complex analysis to R. O. WELLS, JR., Department of Mathematics, Rice University, Houston, TX 77001.

Abstract analysis to WILLIAM B. JOHNSON, Department of Mathematics, Ohio State University, Columbus, OH 43210.

Algebra and number theory (excluding universal algebras) to MICHAEL ARTIN, Department of Mathematics, Room 2-239, Massachusetts Institute of Technology, Cambridge, MA 02139.

Logic, foundations, universal algebras and combinatorics to JAN MYCIELSKI, Department of Mathematics, University of Colorado, Boulder, CO 80309.

Topology to WALTER D. NEUMANN, Department of Mathematics, University of Maryland, College Park, College Park, MD 20742.

Global analysis and differential geometry to ALAN D. WEINSTEIN, Department of Mathematics, University of California, Berkeley, CA 94720.

Probability and statistics to STEVEN OREY, School of Mathematics, University of Minnesota, Minneapolis, MN 55455.

All other communications to the editors should be addressed to the Managing Editor, R. O. WELLS, JR.

MEMOIRS are printed by photo-offset from camera-ready copy fully prepared by the authors. Prospective authors are encouraged to request booklet giving detailed instructions regarding reproduction copy. Write to Editorial Office, American Mathematical Society, P.O. Box 6248, Providence, Rhode Island 02940. For general instructions, see last page of Memoir.

SUBSCRIPTION INFORMATION. The 1982 subscription begins with Number 255 and consists of six mailings, each containing one or more numbers. Subscription prices for 1982 are $80.00 list; $40.00 member. Each number may be ordered separately; *please specify number* when ordering an individual paper. For prices and titles of recently released numbers, refer to the New Publications sections of the **NOTICES** of the American Mathematical Society.

BACK NUMBER INFORMATION. For back issues see the AMS Catalogue of Publications.

TRANSACTIONS of the American Mathematical Society

This journal consists of shorter tracts which are of the same general character as the papers published in the MEMOIRS. The editorial committee is identical with that for the MEMOIRS so that papers intended for publication in this series should be addressed to one of the editors listed above.

Subscriptions and orders for publications of the American Mathematical Society should be addressed to American Mathematical Society, P. O. Box 1571, Annex Station, Providence, R. I. 02901. *All orders must be accompanied by payment.* Other correspondence should be addressed to P. O. Box 6248, Providence, R. I. 02940.

MEMOIRS of the American Mathematical Society (ISSN 0065-9266) is published bimonthly (each volume consisting usually of more than one number) by the American Mathematical Society at 201 Charles Street, Providence, Rhode Island 02904. Second Class postage paid at Providence, Rhode Island 02940. Postmaster: Send address changes to Memoirs of the American Mathematical Society, American Mathematical Society, P. O. Box 6248, Providence, RI 02940.

Table of Contents

Abstract

Let ϕ be a C^r flow on the noncompact surface M. Call ϕ globally C^r structurally stable $(r \geq 1)$ if ϕ has a neighborhood U in the strong (Whitney) C^r topology such that the orbits of every flow $\psi \in U$ can be mapped onto orbits of ϕ by a homeomorphism $h : M \to M$ which is in a pre-assigned compact-open neighborhood of the identity. _Theorem A:_ If ϕ satisfies: (i) there are no non-trivial minimal sets, and no oscillating orbits; (ii) every restpoint and periodic orbit is hyperbolic; (iii) the closures of the sets $W^+(\phi)$ (resp. $W^-(\phi)$) of stable (resp. unstable) separatrices to fixed saddles and saddles at infinity meet only in fixed saddles, then (a) $\Omega(\phi) = Per(\phi)$, and (b) ϕ is globally C^r structurally stable. _Theorem C:_ For flows in the plane, the condition $W^-(\phi) \cap W^+(\phi) = \emptyset$ is C^r-generic. _Theorem B:_ When $M = \mathbb{R}^2$, the hypotheses of theorem A are necessary as well as sufficient for global C^r structural stability.

AMS (1980) Subject Classification: 58F10, 34D30.

Key Words and Phrases:

Structural stability, flows, saddles at infinity, prolongational limits, separatrices, saddle connections, oscillatory orbits, completely unstable flows, genericity.

Library of Congress Cataloging in Publication Data

Kotus, Janina.
  Global structural stability of flows on open surfaces.

  (Memoirs of the American Mathematical Society, ISSN 0065-9266 ; no. 261)
    Includes bibliographical references and index.
    1. Differentiable dynamical systems.
2. Stability. 3. Surfaces. I. Krych, Michał, 1948-   . II. Nitecki, Zbigniew. III. Title.
IV. Series.
QA3.A57 no. 261 [QA614.8]   510s   81-22941
ISBN 0-8218-2261-6   [514'.32]   AACR2

Acknowledgements

The proofs of our main results were first formulated during the last author's visit to Warsaw in summer 1980 under the PAN-NAS Scientific Exchange Program. He (ZN) would like to thank the National Academy of Sciences (U.S.) and the Polish Academy of Sciences for their support, and the Institute of Mathematics, University of Warsaw for its hospitality.

The bulk of the paper was written during the second author's stay at Tufts University during the academic year 1980-81. He (MK) would like to thank Tufts University for their support and the Department of Mathematics for its hospitality.

The last author (ZN) was partially supported by the National Science Foundation under Grant No. MCS-8102122 during the preparation of this paper.

Finally, we would like to thank Mrs. Elizabeth Branson for her beautiful typing of our manuscript.

1. *Statement of results*

This paper concerns the structural stability of flows on surfaces (two-dimensional manifolds) which are not compact.

Recall that the dynamical systems ϕ and ψ are *topologically equivalent* if there exists a homeomorphism h between their phase spaces, taking ϕ-orbits to ψ-orbits and respecting time order. We will refer to h as the *equivalence homeomorphism* between ϕ and ψ. A system is *structurally stable* if it is topologically equivalent to every system sufficiently near it in a suitable topology. The study of structurally stable systems has been a central theme of research in dynamical systems, especially in the last twenty years.

This period has seen the development of a comprehensive theory of cascades and flows on a *closed manifold* (smooth and compact without boundary). In this setting, dynamical systems are topologized by the *uniform C^r topology* (uniform convergence of functions together with their derivatives to order r) applied to the generating map for cascades or the velocity vector field for flows. Usually in structural stability problems one takes $r = 1$. The work of many people [AP, M 1-2, Sm 1-3, DeB, PePe, Pe 1-3, An, Mo, Pu, PS, Pa, PaS, Ro, R 2-4, DeM] has distilled three general conditions for C^1 structural stability of a dynamical system ϕ on a closed manifold M. Associated with ϕ are two invariant subsets of M, the *periodic points*

$$\mathrm{Per}(\phi) = \{x \mid \phi(t,x) = x \text{ for some } t > 0\}$$

and the *nonwandering points*

$$\Omega(\phi) = \{x \mid \exists\, x_n \to x, \ t_n \to \infty \ .\ni.\ \phi(t_n, x_n) \to x\} \ .$$

Roughly stated, the following three conditions are sufficient for C^1 structural stability of ϕ when M is closed:

    (i) $\mathrm{Per}(\phi)$ is dense in $\Omega(\phi)$

    (ii) $\Omega(\phi)$ has hyperbolic structure

    (iii) stable and unstable sets for $\Omega(\phi)$ are in general position.

There is considerable evidence that these conditions are also necessary [F 1-3, G, Ma 1-2, L, Pℓ 1-2, R 1-6].

In this paper we shall concentrate on the formulation of conditions (i)-(iii) as they apply to flows on surfaces; the reader is referred to the literature for the general formulation [Sm 2, Ro, R 2-4,7, Ni 6, Sh, Bo, I].

Received by the editors August 31, 1981.

1

Suppose ϕ is a flow on the surface M . Orbits in $\mathrm{Per}(\phi)$ are of two types: rest points and circular (periodic) orbits. Recall that a *hyperbolic rest point* of ϕ is a point p at which the velocity $\dot{\phi}$ vanishes, while the eigenvalues of its Jacobian matrix $D\dot{\phi}(p)$ are not pure imaginary. If both eigenvalues have negative (resp. positive) real part, then every point near p moves toward p in forward (resp. backward) time, and p is a *fixed sink* (resp. *source*). We shall refer to the set of all points in M which tend toward a sink (resp. source) in forward (resp. backward) time as the *basin of attraction* (resp. *region of repulsion*) of the sink (source). A hyperbolic rest point which is neither a sink nor a source is a *fixed saddle*. In this case, there are two distinguished pairs of orbits, the *stable* (resp. *unstable*) *separatrices* of p, which tend to p in forward (resp. backward) time. All other orbits leave any sufficiently small neighborhood of p in both time directions. A *saddle connection* between fixed saddles p and q (possibly with $p = q$) is an orbit which is simultaneously a stable separatrix for q and an unstable separatrix for p .

A *hyperbolic periodic orbit* for a flow ϕ on a surface is a circular orbit whose characteristic exponent (in the sense of Floquet or Liapunov), defined as the integral of $\mathrm{div}(\phi)$ over the orbit, is nonzero. It is a *periodic sink* (resp. *source*) when the exponent is negative (resp. positive). We define basins of attraction and regions of repulsion by analogy with rest points.

A well-known theorem of Peixoto characterizes the C^1 structurally stable flows on a closed surface by a version of the general conditions mentioned above:

(1.1) <u>Theorem</u> ([Pe], cf [Gu 1-3]):

If M is a closed 2-manifold, then a flow ϕ on M is C^1 structurally stable if and only if the following three conditions hold:

   (i) $\Omega(\phi) = \mathrm{Per}(\phi)$

   (ii)   every orbit in $\mathrm{Per}(\phi)$ is hyperbolic

   (iii)   there are no saddle connections.

Furthermore, these conditions hold for a dense open set of flows on M in the uniform C^1 topology.

By contrast with a closed manifold, an *open manifold* is a smooth finite-dimensional manifold without boundary which is metrizable but not compact. Ignoring the real line, on which flows have a trivial structure, the simplest example of an open manifold is $\mathbb{R}^2$, which arises naturally as the phase space of a second-order differential equation or system. Our results concern the analogue of (1.1) for flows on open surfaces.

We adopt the *strong* (or *Whitney*) C^r *topology* on vector fields, in which a neighborhood of ϕ is defined by global pointwise estimates of the form

$$\| f(x) - g(x) \| < \varepsilon(x)$$

where ε is a continuous strictly positive function on M, and f (resp. g) represent corresponding components of $\dot{\phi}$ (resp. $\dot{\psi}$) and their derivatives to order r in some atlas of local coordinates on M. This topology is independent of our choice of metric and coordinates on M, by contrast with the (weaker) uniform topology (ε constant), which is metric-dependent on an open manifold. We note also that the strong topology is stronger than the coefficient topology used in studying polynomial vector fields [Ch, Go, T].

We call a flow ϕ on M *globally C^r structurally stable* if for every flow ψ in some strong C^r neighborhood of ϕ there exists a homeomorphism $h : M \to M$ which takes ϕ-orbits to ψ-orbits (respecting time order) and is near the identity map on M with respect to the compact-open topology. We will discuss several technical aspects of this definition below, in §2.

It is well known [PePu, TW, Kr 4] that any open manifold of dimension 2 or more supports a nonempty C^1-open set of flows which fail to be structurally stable, by contrast with (1.1) in the closed two-dimensional case. (We outline an example in §2). Our results in this paper show how conditions (1.1 (i)-(iii)) can be extended to encompass behavior "at infinity" so as to give general sufficient conditions for a flow ϕ on any open 2-manifold M to be C^r structurally stable ($r \geq 1$); these conditions are necessary in case $M = \mathbb{R}^2$.

Given a flow ϕ on the open surface M and a point $x \in M$, denote its orbit by $O(x, \phi)$, and define the *positive* (resp. *negative*) *semi orbit* by

$$O_+(x,\phi) = \{\phi(x,t) | t > 0\}$$
$$O_-(x,\phi) = \{\phi(x,t) | t < 0\} .$$

We distinguish three kinds of asymptotic behavior for each semi orbit $O_\pm(x,\phi)$:

(a) $O_\pm(x,\phi)$ is *bounded* if it is contained in some compact set $C \subset M$;

(b) $O_\pm(x,\phi)$ *escapes to infinity* if for each compact set $C \subset M$ there exists a point $y \in O_\pm(x,\phi)$ for which $O_\pm(y,\phi) \cap C = \emptyset$;

(c) $O_\pm(x,\phi)$ *oscillates* if it is neither bounded nor escapes to infinity.

These kinds of behavior for $O_+(x,\phi)$ (resp $O_-(x,\phi)$) can also be described in terms of the ω-*limit* (resp. α-*limit*) set of $x \in M$ under ϕ:

$$\omega(x,\phi) = \{y | \exists t_n \to +\infty \ni \phi(x,t_n) \to y\}$$
$$\alpha(x,\phi) = \{y | \exists t_n \to -\infty \ni \phi(x,t_n \to y\} .$$

It is easy to see that $O_+(x,\phi)$

(a)   is bounded iff $\omega(x,\phi)$ is compact (and nonempty)

(b)   escapes to infinity iff $\omega(x,\phi) = \emptyset$

(c)   oscillates iff $\omega(x,\phi)$ is a noncompact subset of M .

In our formulation of our main theorem, we obtain the condition $\Omega(\phi) = \text{Per}(\phi)$ as a conclusion. This requires a hypothesis on bounded orbits not included in (1.1). Recall that a *minimal set* is a (nonempty) compact invariant set with no proper compact invariant subsets. Trivial minimal sets are rest points and circular orbits; we assume there are no others (it is easy to see from the well-known theorem of Schwartz [Sch] that this condition is necessary for C^r stability of flows on any surface).

We will also assume that unbounded orbits escape to infinity (no oscillation), and formulate a further condition on these orbits in terms of the *prolongational limit sets* of Ura [U] and J. Auslander [Au, AuS]. (An accessible introductory account of Auslander and Ura's theory is given in [BhS]). The role of these sets in the perturbation theory of flows with $\Omega(\phi) = \emptyset$ was previously studied in [Ni 1-5]. The first positive (resp. negative) prolongational limit set of $x \in M$ under ϕ is

$$J^{\pm}(x,\phi) = \{y \mid \exists\, x_n \to x,\ t_n \to \pm\infty \ni \phi(t_n, x_n) \to y\} .$$

In general, $\omega(x,\phi) \subset J^+(x,\phi)$ and $\alpha(x,\phi) \subset J^-(x,\phi)$, and one sees easily that

$$\Omega(\phi) = \{x \mid x \in J^+(x,\phi)\} = \{x \mid x \in J^-(x,\phi)\} .$$

Modifying the definition of Nemytskii-Stepanov [NS], we say two unbounded semi orbits $O_+(x,\phi)$ and $O_-(y,\phi)$ form a *saddle at infinity* if each escapes to infinity and

$$y \in J^+(x,\phi) \ (\text{ie, } x \in J^-(y,\phi)) .$$

In this case, we call $O_+(x,\phi)$ (resp. $O_-(y,\phi)$) the *stable* (resp. *unstable*) *separatrix* of the saddle at infinity.

Given a flow ϕ on the open surface M, denote by $W^+(\phi)$ (resp. $W^-(\phi)$) the union of all stable (resp. unstable) separatrices of saddles - fixed and at infinity. Each set is ϕ-invariant; it may consist of finitely or infinitely many distinct orbits. In either case, it is not generally closed, since a fixed saddle belongs to the closure of each of its separatrices. Furthermore, when $W^+(\phi)$ has infinitely many orbits, it may accumulate on orbits outside $\text{Per}(\phi)$ which are not themselves in $W^+(\phi)$. We shall see examples of these phenomena in §2.

With these definitions, we can formulate sufficient conditions for global structural stability:

(1.2) Theorem A:

If dim M = 2 and ϕ is a C^r flow on M (r $\geq$ 1) satisfying:

    (i) there are no non-trivial minimal sets, and no
           oscillating orbits

    (ii) every orbit in Per(ϕ) is hyperbolic

    (iii) clos $W^-(\phi) \cap$ clos $W^+(\phi) \subset$ Per(ϕ) ,

then

    (a) $\Omega(\phi)$ = Per(ϕ)

    (b) ϕ is globally C^r structurally stable.

We note that (i) and (ii) alone do not imply (a) (see §2). When $M = \mathbb{R}^2$, there are never any non-trivial minimal sets, and for M any open surface standard arguments from the closed case show that existence of either a non-trivial minimal set or a non-hyperbolic periodic point yields structural in-stability of a flow on M . The other conditions of (1.2) are considerably easier to handle when $M = \mathbb{R}^2$ than in general. We have succeeded in showing necessity only in this case:

(1.3) Theorem B:

    If $M = \mathbb{R}^2$ and ϕ is a globally C^r structurally stable flow on M, then (1.2) (i)-(iii) hold.

However, we have also found no counterexamples to (1.3) with M an open surface, and conjecture that it holds with $M = \mathbb{R}^2$ replaced simply by dim M = 2 . To understand the difficulty in this extension, we recall the following theorems (in their two-dimensional version).

(1.4) Kupka-Smale Theorem [Ku, Sm 4, Pe 3]

    On any closed or open surface M , the flows ϕ for which

        (i) every orbit in Per(ϕ) is hyperbolic

        (ii) there are no saddle connections between fixed saddles

are C^r generic for r $\geq$ 1 .

(1.5) Theorem [Ko] (see also (7.6))

    The flows on $\mathbb{R}^2$ with no oscillating orbits are C^r generic for r $\geq$ 1 .

We prove (1.3) by combining (1.4) and (1.5) with the following result, which was proved in the special case $\Omega(\phi) = \emptyset$ in [Kr 1]:

(1.6) Theorem C:

    The flows on $\mathbb{R}^2$ satisfying the conditions of (1.4) and (1.5) and
$$W^-(\phi) \cap W^+(\phi) = \emptyset$$
are C^r generic for r $\geq$ 1 .

The proofs of (1.5) and (1.6) both make extensive use of the Jordan curve theorem and related topological properties of $\mathbb{R}^2$. It can in fact be shown that (1.6) is false if one replaces $\mathbb{R}^2$ with the complement in $\mathbb{R}^2$ of a cantor set (see §2). Furthermore, the analogue of (1.6) with $W^{\pm}(\phi)$ replaced by their closures (and "$= \emptyset$" with "$\subset$ Per(ϕ)") is false on any open surface, as shown by the examples of Peixoto-Pugh [PePu], Takens-White [TW], and Krych [Kr 4], in which $W^{\pm}(\phi)$ are each a dense subset of M (including $M = \mathbb{R}^2$), for a C^r-open set of flows.

Thus, we cannot use the genericity approach to prove theorem B when $\mathbb{R}^2$ is replaced by an arbitrary open surface. Nevertheless, in a manner similar to the proofs of instability in the counter-examples cited above, it may be possible to show that the existence of oscillatory orbits or of common (non-periodic) accumulation points of $W^+(\phi)$ and $W^-(\phi)$ prevents global structural stability.

This situation contrasts with that of Ω-stability, studied in the plane by Klok [Kℓ]. Two dynamical systems are Ω-*equivalent* if there is an equivalence homeomorphism between their nonwandering sets. A flow ϕ is (C^r) Ω-*stable* if, given a neighborhood U of $\Omega(\phi)$, for every strong C^r perturbation ψ of ϕ, (i) each component of U which contains points of $\Omega(\phi)$ also contains points of $\Omega(\psi)$, and (ii) ϕ and ψ are Ω-equivalent. One can also define a notion of *R-stability,* replacing the nonwandering set Ω with the Auslander recurrent set R (see [Kℓ] for definitions). Klok shows that R-stability (resp. Ω-stability) is C^r-generic for flows in $\mathbb{R}^2$, and is equivalent to (resp. implied by) the conditions

(i) $\Omega(\phi) = $ Per(ϕ)

(ii) every orbit in Per(ϕ) is hyperbolic

(iii) there are no generalized cycles

where a "generalized cycle" is a transfinite sequence x_α (α ranging over some ordinal A) such that $x_\alpha \in $ clos $\bigcup_{\beta < \alpha} J^+(x_\beta)$ and $x_0 = x_A \notin $ Per(ϕ). Condition (iii) allows $W^{\pm}(\phi)$ to intersect outside Per(ϕ), but requires that the closed relation obtained naturally from $x < J^+(x)$ is an ordering.

For example, in the following section, examples (2.1) and (2.2) have cycles, while (2.3) and (2.4) do not.

Before proceeding with our proofs of theorems A, B and C, we devote section 2 to a detailed discussion, with examples, of our definition of global structural stability, and of the relation between our results and earlier work.

We have learned recently of independent unpublished work by C. Camacho and R. Mañe, in which results similar to ours have apparently been proved.

2. *Examples and discussion of results*

In this section, we review earlier work concerning structural stability on open manifolds, illustrating with examples the significance of three technical points in our definition of global C^r structural stability: our choice of the strong C^r topology to measure the size of perturbations, the requirement that the equivalence homeomorphism h be globally defined on M , and the requirement that h be near the identity map in the compact-open topology. The first and second of these points are closely inter-related, and can best be understood by comparison with an earlier approach to structural stability.

The original formulation of structural stability by Andronov and Pontriagin [AP] involved a definition of structural stability which simply ignores all behavior outside some compact set. If M is an open manifold, a *compact region* in M is a compact submanifold-with-boundary $G \subset M$ of co-dimension zero. Two flows ϕ, ψ on M are *topologically equivalent on G* if there exists a homeomorphism $h : G \to G$ taking a (directed) ϕ-orbit segment in G to a (directed) ψ-orbit segment in G. Note that this kind of equivalence respects periodicity and the nonwandering property only for orbits entirely contained in G , since different segments of one global ϕ-orbit may map into different global ψ-orbits. A flow ϕ is C^r *structurally stable on G* if any flow on M whose velocity at points of G is uniformly C^r near that of ϕ is topologically equivalent to ϕ on G .

DeBaggis [DeB] published a proof of the statement by Andronov-Pontriagin that when G is a compact region in $\mathbb{R}^2$ and ϕ is a flow transverse to the boundary of G , then conditions (i)-(iii) of theorem (1.1) imply that ϕ is structurally stable in G . The necessity of these conditions was considered by Markus [M2]. The Peixotos [PePe] removed the transversality requirement on ϕ , replacing it with conditions on the kind of internal tangencies an orbit can have with the boundary of G . This approach was extended to higher dimensions by Percell [Per] and Sotomayor [So 2] under simplifying assumptions on the dynamics interior to G . Assuming [Per] that every semi orbit leaves G or [So 2] that the Morse-Smale conditions (a higher-dimensional analogue of (1.1 i-iii)) hold in G, they formulated rather involved tangency conditions at the boundary of a compact region which insure C^1 structural stability on G . Some of the technical complexity of these conditions results from the requirement that the equivalence homeomorphism h map G exactly onto itself.

A variation of this definition (see [ALGM]) requires only that h be an embedding of G in M which is uniformly C^0-near the inclusion. Using the varied definition, Robinson [R 5] formulated simpler boundary conditions, subsuming more complicated internal dynamics in G, which insure that a flow ϕ is structurally stable on G with "moveable boundary."

Sotomayor [So 2] has proposed a global definition of structural stability based on these ideas: he would regard a flow ϕ on M to be structurally stable if it is structurally stable on each of a countable nested family $G_i \subset G_{i+1}$ of compact regions whose union is M. However, data taken from a bounded region cannot detect either oscillatory behavior or saddles at infinity, and this limits the consequences of structural stability on compact regions.

For example, the second-order equation

$$(2.1) \qquad \ddot{x} = a\,\dot{x}(\dot{x} + 1)(\dot{x}^2 + 1)^{-1} - x(\dot{x} + 1)$$

where $0 < a$ is a constant, can be regarded as the phase-plane system

$$(2.1') \qquad \begin{aligned} \dot{x} &= y \\ \dot{y} &= a\,y(y + 1)(y^2 + 1)^{-1} - x(y + 1) \,. \end{aligned}$$

This has a fixed source at the origin, from which oscillatory orbits spiral out toward the line $y + 1 = 0$, which is their ω-limit set (see fig. 2.1a). If one restricts attention only to a bounded region, every orbit eventually goes out of view, and one fails to detect that the line $y + 1 = 0$ is non-wandering. Thus, one can find bounded regions G_i on which (2.1) is C^1 structurally stable by the Peixotos' theorem [PePe]. On the other hand, given one such region G_{i-1} consider two points on the line $y + 1 = 0$ far to the right and left of G_i. It is clear that a downward push very near the right point and a complementary upward push near the left point can be used to create large-amplitude periodic orbits for flows arbitrarily near (2.1') on G (Fig. 2.1b). These orbits are not detected by h, even though they intersect G_i.

A more drastic example of this type is the system on the cylinder $S^1 \times \mathbb{R}$ defined by

$$(2.2) \qquad \begin{aligned} \dot{\theta} &= \sin^2 \theta \\ \dot{z} &= \cos \theta \,. \end{aligned}$$

This global system has no nonwandering orbits whatsoever. However, the pairs of semi-orbits $O_+(0, z_0)$, $O_-(\pi, z_1)$ and $O_+(\pi, z_1)$, $O_-(2\pi, z_0)$ are vertical rays forming two saddles at infinity. (Fig. 2.2a) In particular, $W^+(\phi) \cap W^-(\phi)$ is the union of these two lines: $\theta = \pi$ and $\theta = 0$. In this case, two small pushes in the direction of increasing θ, concentrated near the two points (π, z_0) and $(0, z_1)$, can again create periodic orbits of large amplitude. (Fig. 2.2b) Here, any of the theorems [PePe, Per, So 2]

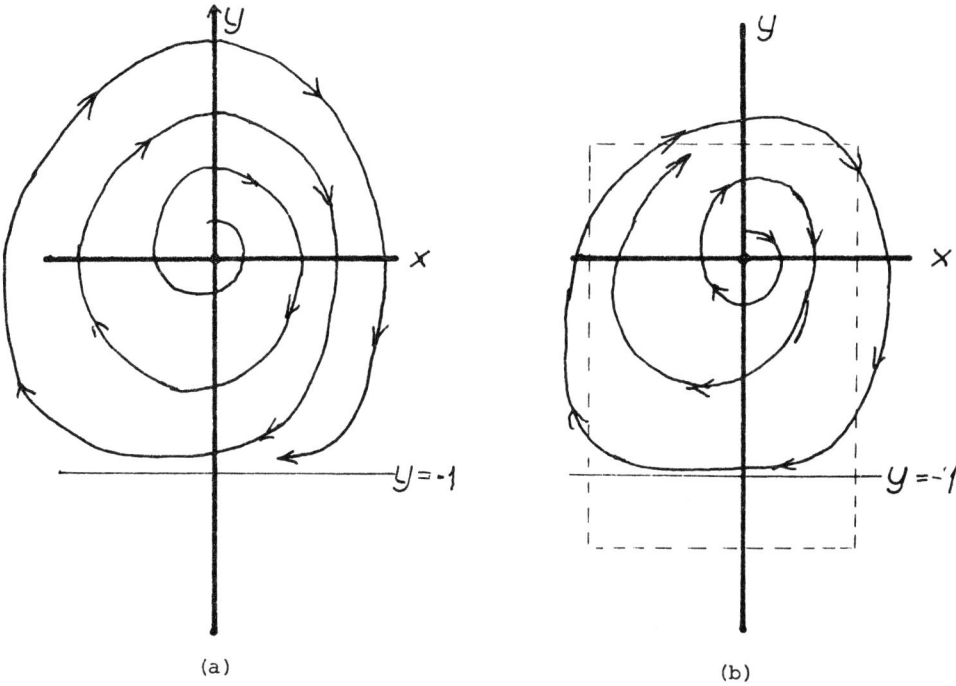

(a)   (b)

Figure 2.1

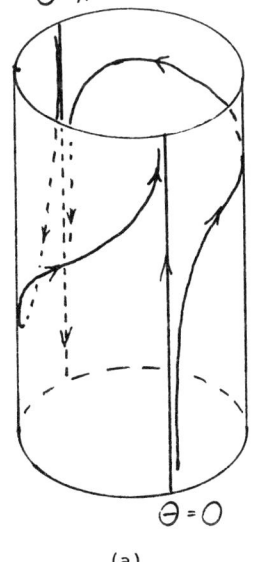

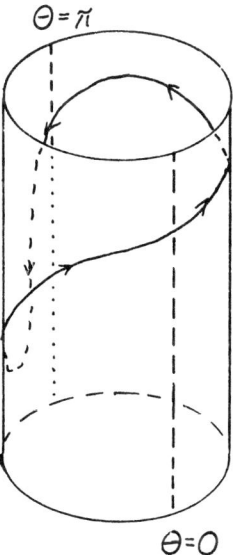

(a)   Figure 2.2   (b)

can be used to show structural stability on a bounded region.

A third example illustrates a somewhat more subtle phenomenon. It can be achieved via the universal covering of (2.2), or by a polynomial vectorfield, for example

(2.3)
$$\dot{x} = x^2 - x^4$$
$$\dot{y} = 4x^2 - 1 .$$

The horizontal component of this vectorfield has simple zeroes along the lines $x = \pm 1$ and a double zero along the y-axis. Each adjacent pair of these lines (moving from left to right) forms a saddle at infinity (Fig. 2.3a) so that

$$W^+(\phi) = \{(x,y)\,|\,x = -1 \text{ or } x = 0\}$$
$$W^-(\phi) = \{(x,y)\,|\,x = 0 \text{ or } x = +1\}$$

and $W^-(\phi) \cap W^+(\phi)$ is the y-axis. A small change in the coefficients of $\dot{x}$ (for example $\dot{x} = x^2 - x^4 + \epsilon$) destroys all prolongational limit relations, but this is a large change in the strong topology. In fact, we will see that saddles at infinity persist under strong perturbation. However, a small push to the right near the origin will pull the two adjacent saddles at infinity in (2.3) apart into two saddles at infinity separated by a parallelizable strip (fig. 2.3b). For a fixed compact region, G , a sufficiently small perturbation will result in a very slight separation between these saddles, and the distinction between figures 2.3a and 2.3b will not be detected inside G .

A variation on this phenomenon is provided by the quadratic vectorfield, shown us by Chicone and Shafer [ChS]

(2.4)
$$\dot{x} = 2xy$$
$$\dot{y} = 2xy - x^2 + y^2 + 1 .$$

It is shown in [ChS] that the flow ϕ of (2.4) is Morse-Smale: $\Omega(\phi)$ consists of the two restpoints $p_\pm = (\pm 1, 0)$; these are, respectively, a source and a sink, and $W^\mp(p_\pm)$ are, respectively, the right and left half-planes

$$W^\mp(p_\pm) = \{(x,y)\,|\,\pm x > 0\} .$$

The y-axis $\gamma = \{(x,y)\,|\,x = 0\}$ is invariant but wandering. However, there are two distinguished orbits $O(q_\pm)$ with $q_\pm \in W^\mp(p_\pm)$ such that $O_+(q_+)$, γ and γ, $O_-(q_-)$ form saddles at infinity (see fig. 2.4). In fact, a small perturbation near the y-axis, pushing to the left, creates orbits with α-limit p_+ and ω-limit p_- . However, for a strong C^1 neighborhood of ϕ , such orbits can be forced to leave any pre-ordained compact set. Thus, again, (2.4) is stable on compact regions but unstable in the global sense.

The notion of structural stability on open manifolds, in the sense which we have adopted, has been considered in several earlier papers under hypotheses

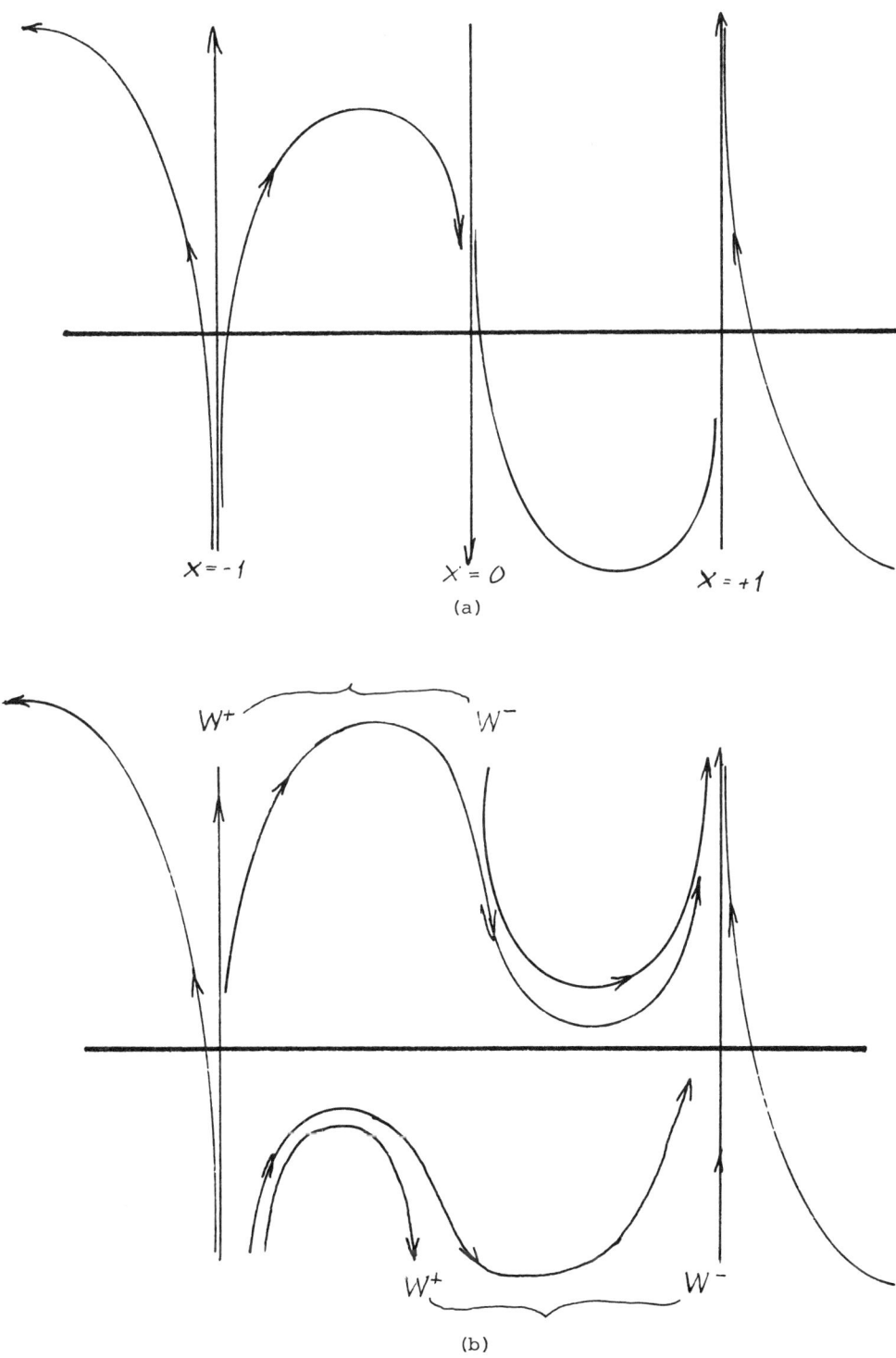

Figure 2.3

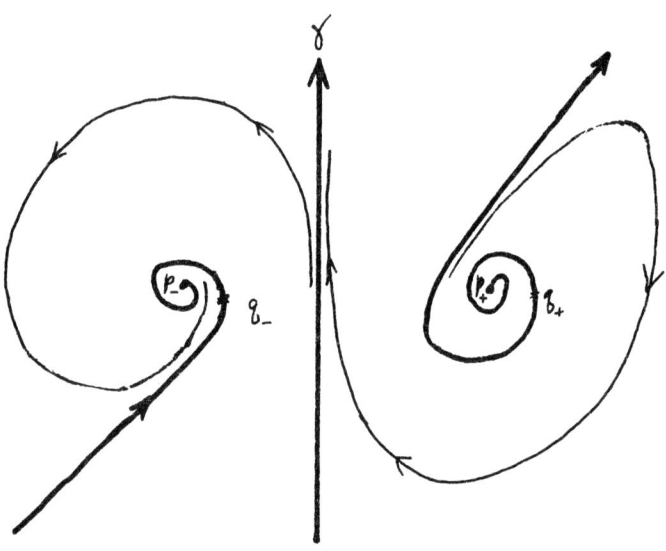

Figure 2.4

that avoid direct consideration of the behavior at infinity. Mendes [Me 1]
has extended to open manifolds the structural stability theorem of Palis and
Smale [Pa, PaS], that a diffeomorphism ϕ for which $\Omega(\phi)$ consists of hyperbolic periodic orbits with stable and unstable manifolds in general position,
and with no saddles at infinity, is structurally stable. On surfaces, his
hypotheses imply those of our theorem A, since the existence of non-trivial
minimal sets or oscillating orbits would also imply non-periodic (or periodic
but non-hyperbolic) orbits in the nonwandering set. The assumption of no
saddles at infinity assures ignorable behavior at infinity, and allows Mendes
to directly adapt to this situation the compact methods of Palis-Smale.

In connection with the hypothesis of ignorable behavior at infinity, we
mention recent work of Nadzieja [Na 1-2], which characterizes dynamical
systems for which every positive semi orbit is bounded and remains bounded
after perturbation.

Another aspect of compact theory which has been adapted to open manifolds
is the structural stability of Anosov systems. Recall that the existence of a
hyperbolic structure (for a dynamical system) on a closed manifold is independent of the metric. This is definitely false in the open case, as demon-
strated by White's [Wh] construction of a complete metric on $\mathbb{R}^2$ for which
a translation has a hyperbolic structure. Mendes [Me 2] has investigated
further the structure of Anosov diffeomorphisms in the plane, and found that

they satisfy the hypotheses of his earlier theorem. This of course is not in general true for Anosov systems on other open manifolds. Osipov [O 1-2], motivated by his geometric regularization of the Kepler problem, has proved the structural stability of Anosov systems in general, but in doing so has made further technical assumptions about the metric in which the system is hyperbolic, which allow him to use the uniform estimates that arise in the compact case. The extent to which the assumption of a global hyperbolic structure under some metric implies intrinsic, metric-independent dynamical properties is not well understood.

We have seen examples of behavior at infinity which prevents global structural stability, and a few results which establish stability when this behavior is trivial. Yet it follows from our theorem A that certain kinds of non-trivial behavior at infinity are consistent with structural stability. For example, the systems

(2.5)
$$\dot{\theta} = \sin\theta$$
$$\dot{z} = \cos\theta$$

on the cylinder $S^1 \times \mathbb{R}^2$ (Fig. 2.5), and

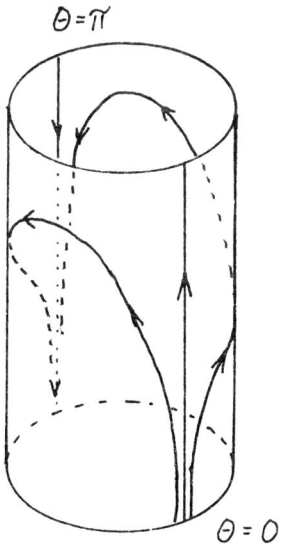

Figure 2.5

(2.6)
$$\dot{x} = x^3 - x$$
$$\dot{y} = 4x^2 - 1$$

in the plane (Fig. 2.6), by contrast with (2.2) and (2.3), have $W^{\pm}(\phi)$ disjoint closed sets, and $\Omega(\phi)$ empty. Therefore they are C^1 structurally

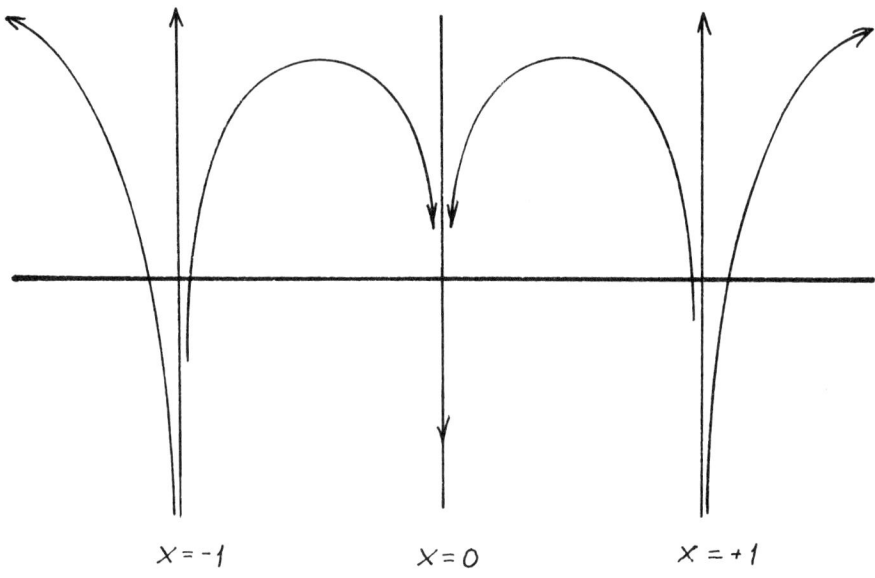

$X = -1$ $X = 0$ $X = +1$

Figure 2.6

stable by theorem A. John Collins [C] first proved the structural stability
of flows like (2.5) in his Ph.D. thesis. He showed that a flow in the plane
for which $\Omega(\phi) = \emptyset$, such that $W^+(\phi)$ and $W^-(\phi)$ are disjoint closed sets of
isolated orbits, are C^1 structurally stable.

The third hypothesis of theorem A - that $W^-(\phi)$ and $W^+(\phi)$ have disjoint closures - is an analogue of the "no saddle connections" hypothesis of
theorem (1.1). The condition that these sets themselves be disjoint has a
crucial simplifying effect on the dynamics. When all orbits in $Per(\phi)$ are
hyperbolic, the condition $W^-(\phi) \cap W^+(\phi)$ has three distinct parts:

1.   No saddle connections

2.   The stable and unstable separatrices of a fixed saddle are
     not involved in any saddle at infinity.

3.   The stable separatrix of a saddle at infinity is not also
     the unstable separatrix of some saddle at infinity.

Even in the absence of saddle connections (ie, for a Kupka-Smale flow), and
even in the plane (where nontrivial minimal sets do not occur), the presence
of generalized saddle connections as in (2) or (3) can lead to relatively
complicated dynamics, for example to $\Omega(\phi) \neq Per(\phi)$. The oscillatory behavior of (2.1') is one such example (here, the line $y + 1 = 0$ forms a
saddle at infinity with itself). A different example is an accumulation of
hyperbolic periodic orbits on a non-periodic orbit, which can be constructed

in the closed case inside a homoclinic contour, and in the plane can also be achieved via accumulation on the separatrices of a fixed saddle (fig. 2.7a) or of a saddle at infinity (fig. 2.7b).

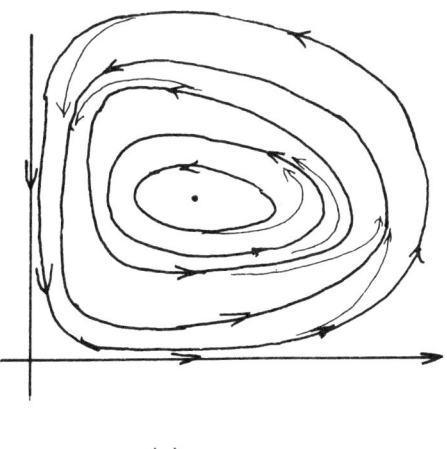

(a)

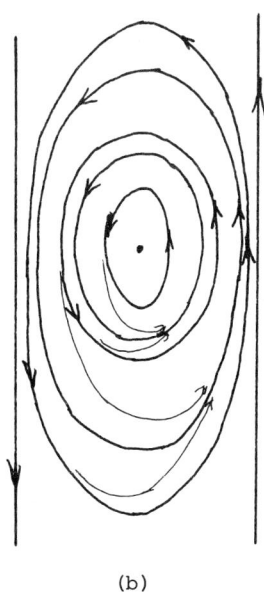

(b)

Figure 2.7

However, as noted in §1, there exist flows ϕ for which orbits in $W^{\pm}(\phi)$ accumulate on non-periodic orbits outside $W^{\pm}(\phi)$ without creating generalized saddle connections. A building block for one class of examples is illustrated by the flow defined on the vertical strip $|x| \leq \pi/2$ by

2.7
$$\dot{x} = f(y - \sec x)\ \cos^2 x$$
$$\dot{y} = 1 + f(y - \sec x)\ [\sin x - 1]$$

where $\sec x = 1/\cos x$ and $f(z)$ is a C^{∞} function that equals 0 for $z \leq 0$, 1 for $z \geq 1$, and is strictly increasing on $(0, 1)$. The flow is sketched in figure 2.8a: the velocity is a constant vertical vector below the curve $y = \sec x$, and has the curves $y = c + \sec x$, $c \geq 1$ as integral curves. Any positive semi orbit which enters the strip $0 \leq y - \sec x \leq 1$ is trapped there, with x strictly increasing along the semi orbit. On the other hand, every point in the region $|x| < \pi/2$, $y < \sec x$ moves straight up until it enters this strip. It follows easily that the flow in the region $|x| < \pi/2$ has one saddle at infinity, formed by the vertical line $x = -\pi/2$ (as stable separatrix) and the curve $y = 1 + \sec x$ (as unstable separatrix). There are no other prolongational relations in this strip.

To construct an example of a flow with $\bar{W}(\phi)$ accumulating on a non-periodic non-separatrix, we place a copy of (2.7) inside each of the vertical bands

$$\frac{1}{n+1} \leq \bar{x} \leq \frac{1}{n}\ , \qquad n = 1, \ldots$$

by conjugation with a diffeomorphism $(x, y) \to (\bar{x}, \bar{y})$ of the form

$$\bar{x} = \frac{2x + (2n+1)\pi}{2\pi n\ (n+1)}$$

$$\bar{y} = y + n\ .$$

Note that the vectorfield so defined is constant vertical on a neighborhood of each of the lines $\bar{x} = 1/n$, $n = 1, \ldots$, and so is C^{∞} on the band $0 < \bar{x} < 1$. Furthermore, since (2.7) is constant vertical for $y < 1$, we see that our new vectorfield is constant vertical below the curve $\bar{y} = 1/\bar{x}$, $\bar{x} > 0$.

For this flow, (fig. 2.8b) we see that the only prolongational relations are between the lines $\bar{x} = 1/n$, and the diffeomorphic images of the curve $y = 1 + \sec x$ (which lie in the region $\bar{y} \geq n + 1$, $\frac{1}{n+1} < \bar{x} < \frac{1}{n}$), for $n = 1, \ldots$. Thus, $\bar{W}(\phi)$ is a closed set, while $W^{+}(\phi)$ consists of the vertical lines $\bar{x} = 1/n$, $n = 1, \ldots$, which accumulate on the $\bar{y}$-axis. In particular, again, there are no nonwandering points. Theorem A applies to show this flow is C^r structurally stable for $r \geq 1$. Using the Whitney approximation theorem [Nr, p. 34] it follows that figure 2.8b can, up to homeomorphism, be exhibited by a real-analytic vectorfield in the plane. In fact, by elaborating this construction, we can obtain structurally stable

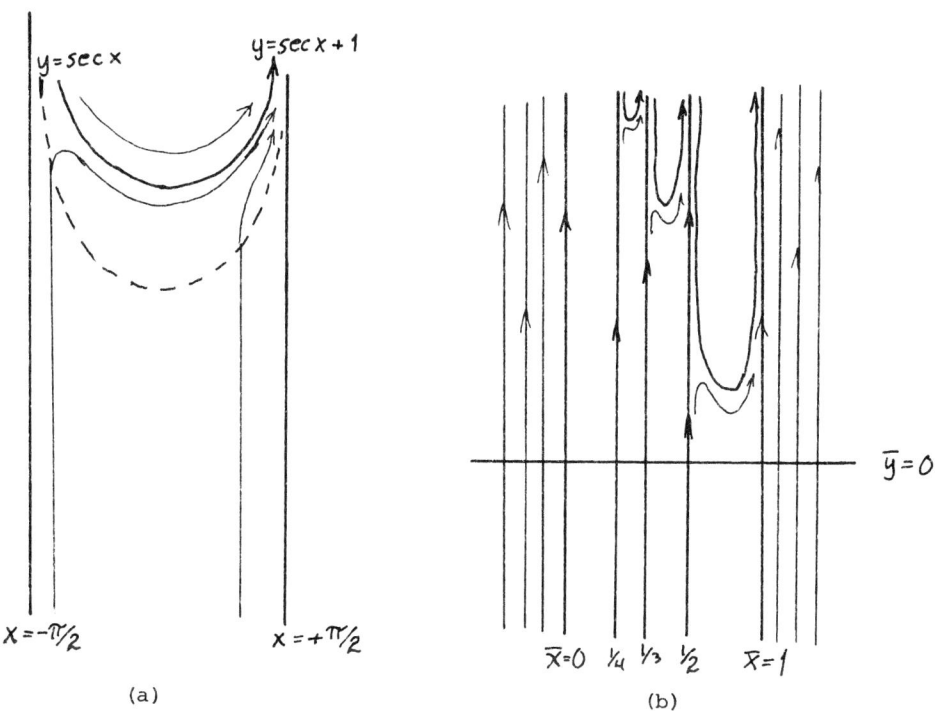

(a)

(b)

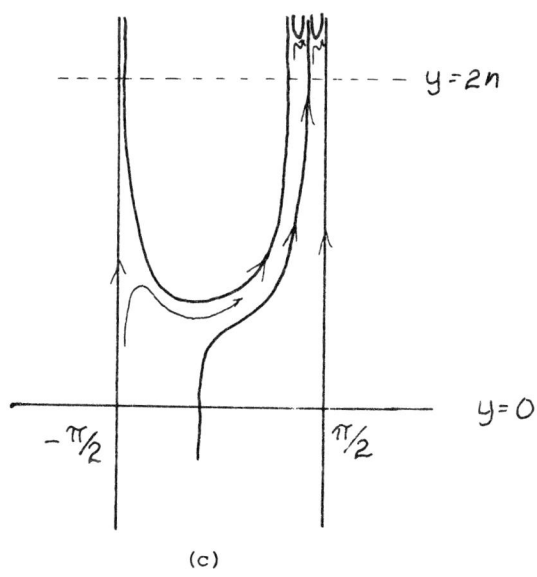

(c)

Figure 2.8

flows ϕ for which clos $W^+(\phi)$ intersects the x-axis in an arbitrary closed, zero-dimensional set.

The various examples of C^1-open sets of non-structurally stable flows [PePu, TW, Kr 4] can be also understood in terms of this example. The flow of fig. 2.8a can, via appropriate diffeomorphisms, be glued into any strip on which the flow has no prolongational relations (ie, is <u>parallelizable</u>). In particular, we can modify fig. 2.8a itself by introducing a copy of this strip into the region bounded by the right half of $y = 1 + \sec x$ and $x = \pi/2$, in such a way that the vectorfield is only changed above the line $x = 2n$, for some $n \gg 1$ (Fig. 2.8c). The effect of this is to increase the intersection of $W^+(\phi)$ with the x-axis, introducing a new point which can be adjusted to lie halfway between the endpoints $x = \pm\pi/2$. Repeating this process, we can modify fig. 2.8c above the line $y = 4n$ so that $W^+(\phi)$ intersects the x-axis at new points halfway between the old, and so on; ultimately, we obtain a flow $\tilde{\phi}$ on the strip $|x| \leq \pi/2$ for which $W^+(\tilde{\phi})$ intersects the x-axis in a dense subset of the interval $[-\pi/2, \pi/2]$. Now, a similar modification below the axis produces a flow $\tilde{\tilde{\phi}}$ for which $W^-(\tilde{\tilde{\phi}})$ and $W^+(\tilde{\tilde{\phi}})$ both intersect $[-\pi/2, \pi/2]$ on the x-axis in a dense set. We can extend this flow to the whole plane by making it π-periodic in x, so that $W^\pm$ each intersect the x-axis in a dense subset. We can also replace the (parallelizable) flow above (and below) the curves corresponding to $y = 1 + \sec x$ of the original flow (2.7) with copies of $\tilde{\tilde{\phi}}$ in such a way that, by induction, we obtain a flow ψ for which $W^\pm(\psi)$ are both dense in $\mathbb{R}^2$. This is essentially the example of Takens-White [TW] and Krych [Kr 4]; the earlier example of Peixoto-Pugh [PePu] constructs a similar phenomenon using stable and unstable separatrices of fixed saddles. In any case, it can be shown that perturbations in the strong C^1 topology do not alter the density of $W^\pm$, and so condition (iii) of theorem A fails on a nonempty C^1 open set of flows. On the other hand, theorem C shows that these flows can be perturbed so that the (unclosed) sets W^+ and W^- are disjoint. The fact that all flows in this open set are structurally unstable is a consequence of the fact that we can pass between disjoint and non-disjoint W^+ and W^-.

It is interesting to note in this connection that theorem C itself fails on certain open surfaces. A simple procedure for obtaining examples of flows with saddles at infinity is to start with a parallel flow (constant velocity), then multiply the velocity by a positive function which vanishes on a set K with nowhere-dense saturation, and finally removing K from the underlying manifold. An integral curve of the original flow which passed through a point $k \in K$ is now separated by K into two or more orbits; if k is isolated in the intersection of K with this curve, then we have two semi orbits of the

new flow escaping in forward (resp. backward) time to the point "at infinity",
k . Since K has nowhere dense saturation, there exist orbits which pass
arbitrarily near both semiorbits and hence form a saddle at infinity.

We apply this procedure to the horizontal flow on $\mathbb{R}^2$ defined by

(2.8) $\dot{x} = 1$

$\dot{y} = 0$

taking as our set K a double copy of a cantor set in the interval, $C \subset [0,1]$,

$$K = K_- \cup K_+$$

$$K_\pm = \{\pm 1/2\} \times C .$$

This yields a flow ϕ on $M = \mathbb{R}^2 \setminus K$ for which

$$W^+(\phi) = \{(x, y) \mid y \varepsilon C, x < 1/2 \text{ and } x \neq -1/2\}$$
$$W^-(\phi) = \{(x, y) \mid y \varepsilon C, x > -1/2 \text{ and } x \neq +1/2\},$$

in particular

$$W^-(\phi) \cap W^+(\phi) = (-1/2, 1/2) \times C .$$

These phenomena can be detected from inside the open surface M by means
of three transverse sections

$$S_t = \{t\} \times [0, 1] , \quad t = \pm 1, 0 .$$

The flow from S_{-1} to S_0 defines a Poincaré map except on the set $S_{-1}^+(\phi)$
of points whose positive semi orbit escapes to a point "at infinity" in K_- .
The image of this Poincaré map is the complement of the set $S_0^-(\phi)$ of points
whose negative semi orbit escapes to infinity at K_- . Similarly, we have a
Poincaré map between S_0 and S_{+1} , except at the sets of points $S_0^+(\phi) \subset S_0$
and $S_{+1}^-(\phi) \subset S_{+1}$ whose semi orbits escape to infinity at K_+ . We note that
for this flow

$$S_{-1}^+(\phi) = \{-1\} \times C$$
$$S_0^+(\phi) = S_0^-(\phi) = \{0\} \times C$$
$$S_{+1}^-(\phi) = \{+1\} \times C$$

and in particular that

$$W^\pm(\phi) \cap S_0 = S_0^\pm(\phi) .$$

Now, pick $C \subset [0, 1]$ to be a cantor set whose Lebesgue measure exceeds
1/2 (say mes $C = 3/4$) and which excludes 0 and 1. Then $S_0^\pm(\phi)$ each have
measure exceeding 1/2. We shall see in (4.8) how to construct "palaces" –
open sets P_{-1}^+, $P_0^\pm$, P_{+1}^- with the property that every negative (resp. positive)
semi orbit in P_i^+ (resp. P_i^-) leaves the palace via S_i, and any positive
(resp. negative) semi orbit entirely contained in P_i^+ (resp. P_j^-) escapes to
infinity at K. For a C^0 perturbation ψ of ϕ , we can define $S_i^+(\psi)$

(resp. $S_j^+(\psi)$ as the set of points in S_i (S_j) whose positive (resp. negative) ψ-semi orbit is contained in P_i^+ (resp. P_j^-). We will prove that for ψ a C^1 perturbation of ϕ, the sets $S_i^\pm(\psi)$ have measure very near that of $S_i^\pm(\phi)$, and in particular

$$\text{mes } S_0^\pm(\psi) > 1/2$$

for ψ C^1-near ϕ. This insures that

$$S_0^-(\psi) \cap S_0^+(\psi) \neq \emptyset$$

and hence the sets $W^-(\psi)$ and $W^+(\psi)$ intersect somewhere in S_0, for all ψ C^1-near ϕ, disproving the analogue of theorem C for $M = \mathbb{R}^2 \setminus K$. Thus, by contrast with saddle connections between fixed saddles (thm. 1.4), connections between saddles at infinity cannot in general be removed.

This example illustrates the fact that the perturbation theory of flows on open manifolds is not a simple analogue of the compact theory, despite the possible impression that our results are simple reformulations of Peixoto's theorem. Another respect in which the compact theory does not translate verbatim is the issue of nearness of the equivalence homeomorphism h to the identity.

Recall from Peixoto's work [Pe 2] that for flows on a closed surface, the requirement that h be C^0-near the identity is redundant in the definition of C^r structural stability ($r \geq 1$). If a flow on M^2 (closed) is equivalent to all its C^r perturbations, then the equivalence in this case can be chosen uniformly near the identity. This is not true in the plane. Our example shows incidentally that without any requirement on nearness of h to the identity, C^r structural stability does not necessarily imply that restpoints be hyperbolic, in contrast to a well-known fact on closed manifolds [Pe 2, M 1-2, F 1, R 1]

Our example is an adaptation of one in [Kr 3], and requires the C^4 topology in place of C^1. Underlying it is the idea of a *composed focus* [So 1 or ALGM, chap. IX]. For our purposes, this is a restpoint which is topologically a sink or a source, which fails to be hyperbolic because the eigenvalues of the linearization are pure imaginary, but which is non-degenerate to the extent that the Poincaré map of the corresponding local flow in polar coordinates has a nonzero third derivative with respect to r at the origin.

(2.7) <u>Lemma</u> [So 1, (3.12)]: *If $r \geq 0$ and a C^r flow ϕ in the disc has a single composed focus as the attractor of all orbits in the disc, then every flow ψ sufficiently C^r-near ϕ is one of two types:*

        (i) topologically equivalent to ϕ (a topological sink)

or   (ii) the unique restpoint is a fixed hyperbolic source encircled by

        an attracting limit cycle which is the ω-limit set of all non-

restpoints in the disc.

The phase portraits for ϕ and its perturbations are sketched in fig. 2.9.
Using (2.8), we can show

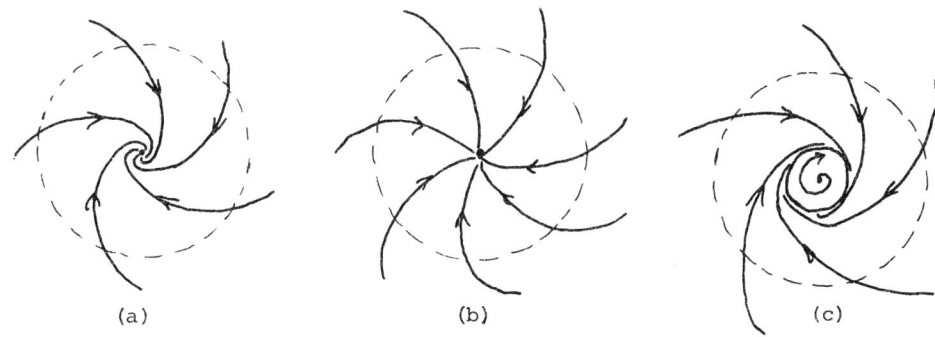

(a)   (b)   (c)

Figure 2.9

(2.9a) <u>*Corollary:*</u> *There exist* c^r *flows* ϕ *on* $\mathbb{R}^2$ *($r \geq 4$) such that* ϕ
is topologically equivalent to any flow ψ *c^r-near* ϕ *, but* ϕ *possesses a*
composed focus.

<u>*Proof*</u>*:*

Our example has restpoints along the x-axis: saddles at $q_n = (n, 0)$
and sinks or sources at $p_n = (n + 1/2, 0)$. The stable separatrices of q_n
form the vertical line $x = n$; these divide the plane into strips
$n < x < n + 1$. All orbits in this strip transversally cross a circle C_n of
radius 1/4 with center p_n . With the exception of the unstable separatrices
of q_n and q_{n+1} , all orbits outside this circle have no α-limit points.
The behavior inside the disc D_n bounded by C_n depends on n . For n > 0,
p_n is a hyperbolic source encircled by a limit cycle that attracts all orbits
(other than p_n) in D_n , and hence in the strip $n < x < n + 1$. For n = 0 ,
we construct the situation assumed in (2.8): p_0 is a composed focus attract-
ing all orbits in D_0 . The phase protrait for $-1 < x < 2$ is sketched in
fig. 2.10.

Now consider a flow ψ c^4-near ϕ . It is easy to see by standard arguments (or the theorem of Mendes [Me 1], or our theorem A) that the orbit
structure of ψ outside D_0 is equivalent to that of ϕ . Inside, (2.8)

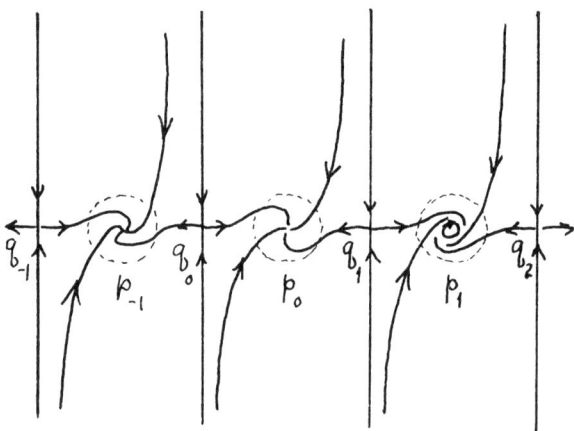

Figure 2.10

tells us that $\phi|D_0$ is equivalent either to $\phi|D_0$ or to $\phi|D_1$. In the first
case, ϕ and ψ are equivalent via a homeomorphism taking each disc D_n to
itself, while in the second case, the equivalence takes the ψ-portrait in D_n
to the ϕ-portrait in D_{n+1} . Of course, in the second case the equivalence
is not near the identity, but in both cases ϕ and ψ are globally equivalent. $\square$

We will see below that this phenomenon does not occur under the C^1
topology. The basic idea is that C^1 perturbations can produce behavior which
is not equivalent to hyperbolic behavior.

Recall that in the compact theory, the behavior of a flow near orbits in
Per(ϕ) can prevent the equivalence homeomorphism from being differentiable.
In a similar way, the behavior of orbits passing near a saddle at infinity can
create an obstruction to differentiability of h , even when $\Omega(\phi) = \emptyset$ (see
[Ne 1]). Thus, the only case in which h can be differentiable is when the
flow is parallelizable [Kr 1,5] (and in this case, the perturbation can be
made in the C^0 topology on C^1 flows [Ni 1]).

Corollary (2.9a) indicates the need for some kind of restraint on the
equivalence to be included in the notion of structural stability. On closed
manifolds, no choice is involved: the uniform C^0 topology is essentially
the only reasonable one. In Osipov's theorem on Anosov systems [O1], there is
a preferred metric, and the equivalence turns out to be near the identity in

the uniform c^0 topology for this metric, for any uniform c^1 perturbation of the system. Thus, one might expect, when we deal with strong c^r perturbations, that the equivalence h should be near the identity in the strong c^0 topology. This requirement, however, is unreasonable, as shown by the following adaptation of an observation in [Kr 1]. Let us, temporarily, say that ϕ is *strongly c^r structurally stable* if given a strong c^0 neighborhood V of the identity, there exists a strong c^r neighborhood U of ϕ such that every $\psi \in U$ is equivalent to ϕ via a homeomorphism $h \in V$.

(2.10) <u>Proposition</u>: *If some orbit of ϕ escapes to infinity in both directions, then ϕ fails to be strongly c^r structurally stable for any r.*

<u>Proof</u>:

Suppose $\alpha(x,\phi) = \emptyset = \omega(x,\phi)$. We construct a neighborhood $V = V_- \cup V_0 \cup V_+$ of $0(x,\phi)$ such that

  (i) V_0 is a flowbox for ϕ containing x

  (ii) $0_\pm(x,\phi)$ is the only ϕ-semi orbit which does not leave $V_\pm$.

(Some technical lemmas which help in carrying out this construction with extra care are given in §4.) Take $\varepsilon : \mathbb{R}^2 \to \mathbb{R}$ a strictly positive function whose value at any point of $0(x,\phi)$ is less than the distance of that point from the boundary of V.

Now suppose ψ is a flow equivalent to ϕ via a homeomorphism h (taking ϕ-orbits to ψ-orbits) such that

  (iii) dist $(h(z), z) < \varepsilon(z)$ for every $z \in V$

and suppose that

  (iv) $\psi(z) = \phi(z)$ for any $z \notin V_0$.

Then we *claim*:

(v) $h[0_+(x,\phi) \cap V_+] \subset 0_+(x,\phi)$.

For suppose $z \in V_+$ belongs to $0_+(x,\phi)$ while $h(z) \notin 0_+(x,\phi)$. By definition of ε, we must have

    $h(z) \in V$

and thus, by the construction of V, and failure of (v),

    $0_+(h(z),\phi)$ leaves V_+.

But

    $0_+(h(z),\phi) = 0_+(h(z),\psi) = h[0_+(z,\phi)]$

by (iv) and the fact that h is an equivalence. But then (iii) implies

    $h[0_+(z,\phi)] \subset V$.

This proves (v), and similar arguments give

(vi) $h[0_-(x,\phi) \cap V_-] \subset 0_-(x,\phi)$.

But (v), (vi) and the fact that h is an equivalence imply that $0_+(x,\phi) \cap V_+$

and $O_-(x,\phi) \cap V_-$ belong to a single ψ-orbit.

But it is easy to find ψ satisfying (iv) arbitrarily c^r-near ϕ for which this last statement is false. Thus, there exist c^r perturbations ψ of ϕ for which no equivalence homeomorphism can satisfy (iii). In other words, ϕ is not strongly c^r-structurally stable. $\square$

Since we are trying as much as possible to formulate our theory in metric-independent terms, the balance between the need for restraints indicated by (2.9) and the impossibility of requiring strong c^0-restraints (2.10) has led us to adopt the version of global structural stability defined in §1, which imposes the compact-open c^0 topology on h .

Before proceeding with a detailed examination of global structural stability, we mention in passing a final minor technicality which results from the fact that our phase space is open: the completeness of vectorfields. A vectorfield with compact support has solution curves defined for all time, but every student of o.d.e. knows examples of differential equations whose solutions have singularities in finite time. However, notice that we are really interested in the orbit structure of the vectorfield, which means that we can reparametrize solution curves. This is easily achieved by multiplying the given vectorfield by a strictly positive function on M .

Thus, suppose we are given an open manifold M . Pick a complete c^∞ Riemannian metric on M . Then a singularity of an integral curve on M can only occur if the curve reaches "infinity" in finite time, which requires its velocity to be unbounded. Thus, given a (smooth) vectorfield X on M , if we multiply it by some positive function $f : M \to \mathbb{R}^2$ so that $\| f(x)X(x) \|$ is globally bounded in M , we obtain a complete vectorfield $f(x)X(x)$. In particular, we can take $f(x)$ to be $(1 + \| X(x) \|^2)^{-1/2}$. It is easy to check that the transformation

$$H : X(x) \to (1 + \| X(x) \|^2)^{-1/2} X(x)$$

is a homeomorphism between the set of all c^r vectorfields on M and the set of c^r vectorfields Y on M satisfying $\| Y(x) \| \le 1$ for all $x \in M$. The latter fields are all complete, and furthermore

(i)   X and $H(X)$ have identical integral curves (with different parametrization)

(ii) X and $H(X)$ have identical sets of hyperbolic restpoints and periodic orbits.

Thus, in any problem where completeness comes into question, we can replace X with its image under H , and the fact that H is a homeomorphism insures that stability of $H(X)$ is equivalent to that of X . Henceforth, we shall entirely ignore this problem.

3. *The structure of well-behaved flows*

This section and the next two are devoted to the proof of theorem A. In this section we will build up a thorough picture of the dynamic structure of a flow satisfying the hypotheses of theorem A. In section 4 we shall establish certain perturbation lemmas. The behavior of certain bounded phenomena is controlled by standard devices from the compact theory, while behavior at infinity is controlled for a single orbit by a variant of Ważewski sets called "towers", which we combine into "palaces" to handle many orbits at once. In section 5, we use palaces to construct a filtration for a flow satisfying our hypotheses, and use this to show that the separatrix structure of such a flow is not changed under C^1 strong perturbation. This allows us to use the method of D. Neumann [Ne 2] to construct an equivalence between a flow and its perturbations, guaranteeing that is is near the identity in the compact-open topology.

We recall the hypotheses of theorem A, which for this section and section 5 serves as a

(3.1) Standing assumption:

M *is a (connected) open 2-manifold, and X is a complete C^1 vector-field on M whose flow ϕ satisfies:*

    (i) there are no non-trivial minimal sets and no oscillating orbits

    (ii) every orbit in $Per(\phi)$ is hyperbolic

    (iii) clos $W^-(\phi) \cap$ clos $W^+(\phi) \subset Per(\phi)$.

We note that since C^1-stability implies C^r-stability for all $r \geq 1$, the fact that we consider C^1 in place of general C^r in (3.1) has no importance. Also, we note that when a flow is *completely unstable* (ie, $\Omega(\phi) = \emptyset$), conditions (i) and (ii) hold vacuously, and in this case (iii) becomes the condition that $W^+(\phi)$ have disjoint closures.

In the first part of this section, we shall prove that (3.1) implies certain dynamic properties, which we summarize in

(3.2) Proposition: *If ϕ satisfies (3.1), then*

    a) $\Omega(\phi) = Per(\phi)$

    b) Any compact set intersects at most finitely many nonwandering orbits.

c) The α- and ω-limit sets of any single point are each either
    empty or consist of a single orbit in Per (φ) .

d) The only prolongational relations between points outside Per(φ)
    are via saddles.

e) Each of the separatrix sets $\overset{+}{W^{-}}(\phi)$ is nowhere dense.

In the second part we shall discuss the structure of canonical and semi-
canonical regions of φ , which will require some new definitions.

To prove (3.2), we run through a sequence of easy technical lemmas concerning the structure of limit sets for flows satisfying (3.1). We will state
these lemmas with a time bias: each one has a backward time version obtained
by replacing ω with α , J^{+} with J^{-} , and "sink" with "source". We leave
the formulation of the backward versions to the reader.

(3.3) <u>Lemma</u>: If ω(x) ≠ ∅ , then it contains an orbit γ in Per(φ) . If
γ is a sink, then $J^{+}(x) = \omega(x) = \gamma$, so that if x is nonwandering, either
x ε γ or γ is a fixed saddle.

<u>Proof</u>:

Since there are no oscillating orbits, ω(x) is empty or compact. If
compact, it contains a minimal set, which must be trivial, hence in Per(φ) ,
and hence is the required γ . If γ is a sink, it uniformly attracts a
whole neighborhood of x , and hence

$$\gamma \subset \omega(x) \subset J^{+}(x) \subset \gamma .$$

In particular, if γ is a sink in ω(x) and x ∉ γ , then x
wanders. □

(3.4) <u>Remark</u>: For any flow, $J^{+}(x)$ cannot contain a source which does not
itself contain x .

(3.5) <u>Lemma</u>: Suppose x ≠ γ , where γ is a fixed saddle and ω(x)
(resp. $J^{+}(x)$) contains γ or one of its separatrices. Then:

    (i) $J^{+}(x)$ contains at least one unstable separatrix of γ

    (ii) if x ∉ $W^{+}(\gamma)$ then ω(x) (resp. $J^{+}(x)$) contains at least
          one stable and one unstable separatrix of γ .

<u>Proof</u>:

Construct a small quadrilaleral Q containing γ and not x whose
edges are transverse to the stable and unstable separatrices of γ (fig. 3.1).
Orbits enter Q via the transversals to $W^{+}(\gamma)$ and leave via the transversals to $W^{-}(\gamma)$. An orbit distinct from the stable separatrices which
enters Q must subsequently leave it. This means that orbits near x which
subsequently come near $W^{+}(\gamma)$ or γ or $W^{-}(\gamma)$ must come near all three, in

particular must come near at least one stable and one unstable separatrix of
γ .   Conclusions (i) and (ii) both follow easily. ☐

(3.6) <u>*Lemma:*</u> *Each unstable separatrix of a fixed saddle either escapes to*
infinity or tends toward a sink in forward time.

<u>*Proof:*</u>

   If $\overline{W}^-(\gamma)$ has a nonempty ω-limit set not equal to a sink, then by (3.3)
it contains a saddle, δ . But then by (3.5),
$$W^+(\delta) \cap \text{clos } W^-(\gamma) \neq \emptyset$$
contradicting (3,1) (iii). ☐

   In the following lemma, remember that $\omega(x) \subset J^+(x)$.

(3.7) <u>*Lemma:*</u>

   (i) *If* $\omega(x) \neq \emptyset$, *then* $\omega(x)$ *consists of a single orbit*
        in $Per(\phi)$.

   (ii) *If* $0(x)$ *is not a source and* $J^+(x)$ *contains a saddle,*
        γ , *then* $x \in W^+(\gamma)$.

<u>*Proof:*</u>

   If $\omega(x)$ contains a sink, there is nothing to prove. If $x \notin W^+(\gamma)$
where γ is a saddle in $J^+(x)$, then some stable separatrix $\sigma \in W^+(\gamma)$
belongs to $J^+(x)$. Since $\alpha(\sigma) \subset J^+(x)$, σ must escape to infinity in backward time. If $\omega(x) = \emptyset$, then $0_+(x)$ and σ form a saddle at infinity,
while if $\omega(x) \neq \emptyset$ it contains a saddle whose unstable separatrix again forms
a saddle at infinity with σ . Either case contradicts (3.1 iii). Thus,
$\gamma \in J^+(x)$ implies $x \in W^+(\gamma)$, and in particular $\omega(x) = \gamma$. ☐

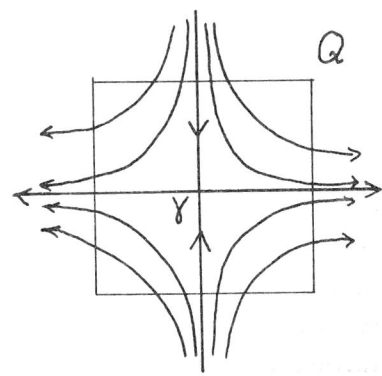

Figure 3.1

Proof of (3.2):

To prove assertion (3.2a), that $\Omega(\phi) = \text{Per}(\phi)$, suppose $x \in \Omega(\phi) \setminus \text{Per}(\phi)$. We know that $x \in J^{\pm}(x)$. If $\alpha(x) = \omega(x) = \emptyset$, then $O_+(x)$ and $O_-(x)$ form a saddle at infinity, so that $x \in W^-(\phi) \cap W^+(\phi)$. If $\omega(x) \neq \emptyset$, then by (3.7) $x \in W^+(\gamma)$ for some fixed saddle γ; since $x \notin W^-(\gamma)$ and $J^-(x)$ intersects $W^+(\gamma)$, some unstable separatrices of γ belongs to $J^-(x)$. Then by (3.6) $O_-(x)$ and this unstable separatrix of γ escape to infinity and hence form a saddle at infinity contrary to (3.1 iii). A similar argument eliminates $\alpha(x) \neq \emptyset$.

Now to see (3.2b), note that an accumulation point of periodic orbits must belong to $\Omega(\phi)$ and hence, by (3.2a), to $\text{Per}(\phi)$. Fixed and periodic sinks and sources are isolated in $\text{Per}(\phi)$, so that this accumulation point must be a fixed saddle. But then, by arguments as in (3.5), these periodic orbits must also accumulate on at least one stable and one unstable separatrix of the fixed saddle. This contradicts (3.1 iii).

(3.2c) is just (3.7i).

To prove (3.2d) suppose $y \in J^+(x)$ and $x, y \notin \text{Per}(\phi)$. We wish to show that $O_+(x)$ and $O_-(y)$ are a stable and unstable separatrix for some saddle (fixed or at infinity). If $\alpha(y) = \emptyset$ then $O_-(y)$ escapes to infinity and $J^-(y)$ contains $\omega(x)$. If $\omega(x)$ is also empty, we are done; if not, then $\omega(x)$ is a fixed saddle γ and $x \in W^+(\gamma)$ (by (3.3) and (3.7)). But then $J^-(y)$ contains an unstable separatrix of γ, which must form a saddle at infinity with $O_-(y)$, contradicting (3.1 (iii)). A similar argument shows that $\omega(x)$ cannot be empty. But then by (3.7 (ii)) $y \in W^-(\gamma)$ and $x \in W^+(\gamma)$ for a saddle $\gamma = \omega(x) = \alpha(y)$.

Finally, we prove (3.2e) in two steps. First, if $x \in W^+(\gamma)$ for some fixed saddle γ, we claim some neighborhood of x intersects $W^+(\phi)$ only in $O(x)$. To see this, consider a quadrilateral Q about γ as in the proof of (3.5); if $x_n \to x$, $x_n \in W^+(\phi) \setminus O(x)$, then $O_+(x_n)$ enters (and hence leaves) Q. This means there exist points $y_n \in O_+(x_n)$ with y_n tending toward an unstable separatrix of γ. This says $W^-(\phi) \cap \text{clos } W^+(\phi) \neq \emptyset$, contrary to (3.1 iii). Thus, the set of stable separatrices to fixed saddles is nowhere dense. Next, we show that the set $W^+_\infty(\phi)$ of stable separatrices to saddles at infinity is nowhere dense. For suppose x is interior to $\text{clos } W^+_\infty(\phi)$. We can assume in fact that $x \in W^+_\infty(\phi)$, and pick y such that $O_+(x)$ and $O_-(y)$ form a saddle at infinity. Since $y \in W^-(\phi)$, it has a neighborhood disjoint from the invariant closed set $\text{clos } W^+(\phi)$. On the other hand, y is a limit of points $y_n \in O_+(x_n)$, where $x_n \to x$, and so x_n (hence y_n) belong to $\text{clos } W^+(\phi)$ for large n, a contradiction. Thus, $W^+_\infty(\phi)$ is nowhere dense, as is its complement in $W^+(\phi)$; this shows $W^+(\phi)$ is nowhere dense. The argument for $W^-(\phi)$ is similar. $\square$

The effect of (3.2e) is, of course, to eliminate the kind of behavior exhibited by example (2.7c) and [PePu, TW, Kr 4].

We turn now to the asymptotic behavior of orbits near a given one. The usual notion behind the word "separatrix" is that such an orbit separates different kinds of asymptotic behavior, so that one expects the asymptotic behavior of all orbits near a non-separatrix to be the same. This means either that a whole neighborhood of points has a common ω- or α-limit set, or that all points of the neighborhood tend toward the same "point of infinity". An intrinsic formulation of this latter idea is the notion of uniform escape to infinity. We say that a set S *excapes to infinity uniformly* in forward (resp. backward) time if for each compact set $K \subset M$ there exists T such that

$$\phi(t,S) \cap K = \emptyset \quad \text{for all} \quad t \geq T \quad (\text{resp. } t \leq T) \ .$$

The hypothesis of Mendes' structural stability theorem [Me 1] which we formulated as nonexistence of saddles at infinity can also be formulated as: every point with unbounded semi orbit has a compact neighborhood which escapes to infinity uniformly. We will show that in our situation a similar statement is true off the separatrix set.

(3.8) Lemma: Suppose C and K are compact sets, and ϕ is a flow satisfying (3.1) for which one of the following conditions holds:
either (i) $C \subset clos \ W^+(\phi)$ and K contains no fixed saddles, or
(ii) C and K are both disjoint from $\Omega(\phi)$, and
$K \cap W^-(\phi) = \emptyset$.
Then there exists T such that

$$\phi(t,C) \cap K = \emptyset \quad \text{for all} \quad t \geq T \ .$$

Proof:

If the conclusion fails, there exist $x_n \in C$, $t_n \to +\infty$, such that $y_n = \phi(t_n, x_n) \in K$. By compactness, we can assume $x_n \to x \in C$ and $y_n \to y \in K$. Then $y \in J^+(x)$. In the first case, we then have $y \in clos \ W^+(\phi) \cap W^-(\phi)$, while in the second we have $y \in W^-(\phi) \cap K$, by (3.2d), in either case a contradiction. $\square$

By a *transverse section* to a flow ϕ, we mean an embedded interval or circle nowhere tangent to the flow: we shall assume that at each end of an interval the embedding either escapes to infinity or is extendible beyond the endpoint as a transverse embedding. Note that we assume all transverse sections are connected, and ipso facto contain no restpoints.

(3.9) *Proposition:*

   If S is a transverse section disjoint from $W^+(\phi)$ and from the periodic sources, then one of the following is true:

              (i) S is contained in the basin of attraction of some sink
                  (fixed or periodic)

or   *(ii)* every compact subset of S escapes to infinity uniformly
                  in forward time.

We note, of course, that (3.9) (like (3.8)) also has a time-reversed version,
which we leave to the reader.

Proof:

   Note that if any point S has a nonempty ω-limit set, that set must (by
3.2c and because $S \cap W^+(\phi) = \emptyset$) be a single sink (fixed or periodic). Suppose γ is a sink whose basin of attraction meets but does not contain S .
γ has a neighborhood contained in its basin $W^+(\gamma)$ and disjoint from S
which is a disc (if γ is fixed), an annulus (if γ is periodic and two-
sided) or a Möbius band. Denote by C the circle or circles forming the
boundary of this neighborhood in M .

   Pick p on the boundary of $S \cap W^+(\gamma)$ and $p_n \to p$, $p_n \in S \cap W^+(\gamma)$.
Let $q_n = \phi(t_n, p_n)$, $t_n > 0$, be the first point of $O_+(p_n)$ on C . We can
assume $q_n \to q \in C$ and (since $p \notin W^+(\gamma)$) $t_n \to \infty$. Thus, $q \in J^+(p)$, and
this contradicts $p \notin W^+(\phi)$.

   Hence, if (i) fails, every point of S escapes to infinity. We wish to
show this happens uniformly on compacta in S . Suppose there exist
$x_i \in S$, $t_i \to \infty$ such that $x_i \in C \subset S$ and $y_i = \phi(t_i, x_i) \in K \subset M$ for some
compact sets C, K . We can assume $x_i \to x \in S$, $y_i \to y \in K$. Thus
$y \in J^+(x)$. By assumption, x escapes to infinity, so by (3.2d) $O_+(x)$
must form a saddle at infinity with some orbit, contradicting
$S \cap W^+(\phi) = \emptyset$. $\square$

   We note that, when S is not compact, the escape to infinity need not be
uniform on all of S (see fig. 3.2).

   These observations allow us to obtain some information concerning the
basin of attraction $W^+(\gamma)$ of a fixed or periodic sink, γ . Denote by
$\partial W^+(\gamma)$ the boundary of the basin (noting that $W^+(\gamma)$ itself is open).

(3.10) *Lemma:*

   If γ is a sink, then for every $x \in \partial W^+(\gamma)$ there exists an orbit
$O(y) \subset W^+(\gamma) \setminus \gamma$ such that $y \in J^+(x)$.

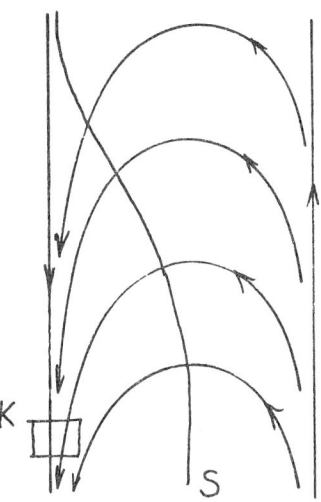

Figure 3.2

Proof:

Since $\overset{+}{W}(\gamma)$ is open, $x \notin \overset{+}{W}(\gamma)$, but there exist $x_n \varepsilon \overset{+}{W}(\gamma)$ with
$x_n \to x$. Pick a compact neighborhood of γ (as in the previous proof) with
boundary C . Since $\gamma = \omega(x_n)$, there exist $t_n > 0$ with
$y_n = \phi(t_n, x_n)\varepsilon \ C$. Since $x \notin \overset{+}{W}(\gamma)$, $t_n \to \infty$, and we can assume $y_n \to y \ \varepsilon \ C$,
so that $y \neq \gamma$ and $y \ \varepsilon \ \overset{+}{W}(\gamma) \cap \overset{+}{J}(x)$. $\square$

An enumeration of the possibilities for x , in light of (3.2d), yields
the following:

(3.11) Corollary:

*If γ is a sink for ϕ satisfying (3.1), then every orbit in $\partial \overset{+}{W}(\gamma)$
is one of five types:*

     (i)   a fixed source

     (ii)   a periodic source

     *(iii)   a fixed saddle with at least one unstable separatrix
          contained in $\overset{+}{W}(\gamma)$*

     (iv)   a stable separatrix for some fixed saddle as in (iii)

     *(v)   an orbit $O(x) \subset \partial \overset{+}{W}(\gamma)$ such that $O_+(x)$, $O_-(y)$
          form a saddle at infinity, where $y \ \varepsilon \ \overset{+}{W}(\gamma) \setminus \gamma$.*

(3.12) Lemma:

If γ is a sink and A is a component of $\overset{+}{W}(\gamma) \setminus \gamma$, then

      (i)   if some semi orbit $O_-(y)$, $y \in A$ excapes to infinity
           in backward time, then either every point in A escapes
           to infinity in backward time or some semi orbit $O_-(z)$,
           $z \in A$, is the unstable separatrix of a saddle whose
           stable separatrices belong to ∂A .

     (ii) if every orbit in ∂A is fixed or periodic, then ∂A
            consists of γ and at most one other orbit.

Remark: If ∂A contains separatrices, then it can also contain several periodic sources: we sketch an example in fig. 3.3, where $\partial W^+(q)$ contains two
periodic sources, two saddles (p_1, p_2) and four separatrices $(W^+(p_1)$,
$W^+(p_2))$.

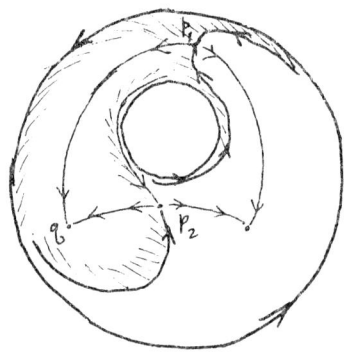

Figure 3.3

Proof of (3.12):

    Pick a compact neighborhood of γ in $W^+(\gamma)$ with boundary transverse to
the flow, as in the preceding two proofs. The part of this boundary in A
consists of a circle S transverse to the flow.

    If $z \in S \cap W^-(\phi)$, then there exists $x \in W^+(\phi)$ with $z \in J^+(x)$. Since
by (3.3) $\gamma \notin \omega(x)$, we must have $x \in \partial A$. If S is disjoint from $W^-(\phi)$,
we can apply the time reversed version of (3.9) to conclude that either S
escapes to infinity uniformly in backward time or else some source σ contains
all of S in its region of repulsion. Since every orbit in A crosses S ,
this would imply that A is contained in the region of repulsion of σ . In
particular, either we have the situation of statement (i), or there is a single
source in ∂A , proving (ii). $\square$

    The basin of attraction of a sink is one example of a region in which
asymptotic behavior in forward time is uniform: we have seen in (3.11) and

(3.12) that there are also some limitations on the variety of backward behavior that can coexist in such a region. We turn now to a description of
maximal regions in which both behaviors are uniform. Such regions have been
studied in a more general context by W. Kaplan [Ka], F. Pluvinage [Plu],
L. Markus [M 1], and D. Neumann and T. O'Brien [NeO]. We adapt the termin-
ology of the latter, and mean by a *canonical region* (for a flow ϕ on M
satisfying (3.1)) a component of the complement of

$$\Omega(\phi) \cup \text{clos } W^-(\phi) \cup \text{clos } W^+(\phi) .$$

The following lemma amplifies an observation made in the plane by Collins [C],
but implicit already in Kaplan [Ka]:

(3.13) Proposition:

   The boundary of a canonical region A   (for a flow satisfying (3.1))
contains:

>   *(i)   at most one sink and one source (fixed or periodic)*
>   *(ii) at most two saddles*
>   *(iii) at most four wandering orbits.*

Proof:

   Note first that A itself is an open connected surface on which the restriction of the flow has no prolongational limits and hence is parallelizable
([NS, M 1,3]). This means that in particular A contains a *cross-section*:
a transverse section S which intersects each orbit in A once. S is either
a line (immersed open interval) or a circle.

   To prove (i), suppose ∂A contains a sink, σ . Since all orbits near σ
and hence some orbits in A belong to the basin of attraction of σ ,   (3.9)
implies all of S is contained in this basin. Thus ∂A has no other sinks.
Similarly, ∂A has at most one source, which if it exists is the common
α-limit of all points in A .

   We will prove (ii) and (iii) together, using the fact that if a saddle
p belongs to the boundary of a canonical region, so do at least one stable
and one unstable separatrix of p. By (3.2a), these are necessarily wandering.

   Thus, suppose ∂A contains a wandering orbit $O(x)$. There exists a
wandering neighborhood of x , hence a transversal $\widetilde{T}$ through x intersecting
any orbit at most once. We restrict attention to a component T of $\widetilde{T} \cap A$,
which we can assume to be a compact interval with one endpoint in A and the
other at x. The flow maps T homeomorphically into the cross-section S
of A , and since T can have only one endpoint in S , S must be a line,
and T maps onto a half-line in S.

   If $y \notin O(x)$ belongs to a second wandering orbit in ∂A , a similar
transversal T' at y maps onto a half line in S . If these half lines are

on the same side of S , then $x \in J^+(y)$, so that $O(x)$ and $O(y)$ are the
stable and unstable separatrices of some saddle (fixed or at infinity).   Thus,
a *third* wandering orbit $O(z) \subset \partial A$ cannot have a transversal   T" *also*
mapping to this side of S .   It follows that there are at most four wandering
orbits in ∂A ,   and two out of any three of these must be prolongationally
related.   This proves (iii) and hence (ii).   □

The palace constructions in §4 will involve the larger sets, which we
call positive (resp. negative) *semi-canonical regions*, defined as components
of the complement of

$$\Omega(\phi) \cup \text{clos } W^+(\phi)   \quad (\text{resp.}   \Omega(\phi) \cup \text{clos } W^-(\phi)).$$

For these regions, (3.13) fails (for example fig. 3.4 shows how a positive
semi-canonical region with many unstable separatrices can contain many sources

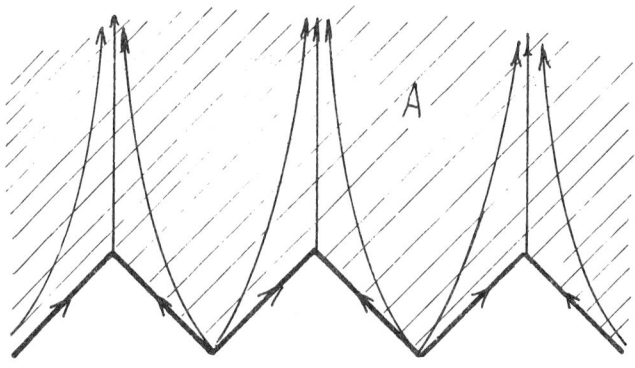

Figure 3.4

and stable separatrices in its boundary).   Nevertheless, something can be said.
To formulate our statement, we define a few terms.   First, noting that separatrices come in pairs (given $x \in W^+(\phi)$ there exists at least one
$y \in W^-(\phi)$ with $y \in J^+(x)$ and vice-versa), we define, given a set $T \subset M$,

$$\overline{W}^+(T,\phi) = \{y \in \overline{W}^+(\phi) \mid y \in J^+(x)   \text{ for some }   x \in T\} .$$

Next, we define the positive, negative and full *saturation* of a set $T \subset M$ by
a flow ϕ as

$$\text{Sat}_\pm(T,\phi) =   \bigcup\{O_\pm(x) \mid x \in T\}$$

$$\text{Sat}(T,\phi) = \text{Sat}_+(T,\phi)   \cup \text{Sat}_-(T,\phi) .$$

(3.14) *Lemma:*

For any compact transverse section, T,

$$clos\ Sat_+(T,\phi) = Sat_+(T,\phi) \cup \overline{W}^-(T,\phi) \cup \bigcup \{\omega(x) \mid x \in T\}$$
$$clos\ Sat_-(T,\phi) = Sat_-(T,\phi) \cup W^+(T,\phi) \cup \bigcup \{\alpha(x) \mid x \in T\}\ .$$

Proof:

For forward time, pick $x_n \in T$, $t_n > 0$ and $y_n = \phi(t_n, y_n) \to y \in clos\ Sat_+(T,\phi)$. By compactness of T, assume $x_n \to x \in T$. If $y \in \mathcal{O}_+(x,\phi)$, then $y \in Sat_+(T,\phi)$. If not, then $t_n \to +\infty$ and $y \in J^+(x)$.

Suppose first that $x \notin Per(\phi)$. By (3.2), either $y \in \omega(x)$ or else x and y belong, respectively, to the stable and unstable separatrices of some saddle, so that $y \in \overline{W}^-(T,\phi)$.

Now, if x is a periodic source, either $y \in \omega(x_n)$ for n large, or $x \in \alpha(y)$ so that $\mathcal{O}_-(y)$ crosses T and $y \in Sat_+(T,\phi)$, or (by 3.11(v)) we can assume $\mathcal{O}_-(y)$, $\mathcal{O}_+(x_n)$ form a saddle at infinity for n large. Thus, $y \in \overline{W}^-(T,\phi)$. □

For a compact transversal, (3.14) says that the boundary of its positive saturation consists of the section, the orbits of its endpoints, ω-limit points, and unstable separatrices for stable separatrices crossing the transversal. For a semi-canonical region, we would like to formulate a similar characterization. Recall from the proof of (3.13) that a *cross-section* for a set A is a transversal S such that $Sat(S,\phi) = A$ and no orbit crosses S twice. When A is a positive (resp. negative) semi-canonical region, we call a cross-section S for A *good* if the boundary of its positive (resp. negative) saturation contains S, at most one sink (source), and at most two semi orbits in $clos\ W^+(\phi)$ ($clos\ \overline{W}^-(\phi)$).

(3.15) *Proposition:*

Suppose A is a positive semi-canonical region. Then there exists a good cross section S to A, and if the boundary of $Sat_+(S,\phi)$ contains a sink, it is the common ω-limit of all points in A.

Proof:

As the proof of (3.13), the flow restricted to the open manifold A is parallelizable and hence has a cross-section S. As a subset of A, S is either a circle or a closed embedding of a line, but in the latter case S may not be nicely embedded in M. However, one easily pulls the "ends" of S away from restpoints and periodic orbits by replacing any part of S inside a prearranged neighborhood of $Per(\phi)$ with part of the boundary of that neighborhood, and replacing an end of S which accumulates at $x \in \partial A \setminus \Omega(\phi)$ with an end of a prearranged one-sided transversal at x.

If $\partial \, \mathrm{Sat}_+(S,\phi)$ contains a sink, then by (3.9) all of S (hence A) is in its basin of attraction.

Suppose now that x is a wandering point in $\partial \, \mathrm{Sat}_+(S,\phi)$; again assume $x \in \partial A$, and take a one-sided transversal T in A with endpoint x. Note that $x \not\in \mathrm{clos} \, \overline{W}(\phi)$. Any compact piece of $T \setminus \{x\}$ maps homeomorphically into S by the flow. All of T must therefore map onto a half-line in S, since either the endpoint of S belongs to $\mathcal{O}_-(x)$ or S escapes to infinity. As in (3.13), two such points x with transversals mapping onto the same end of S must be prolongationally related, so one of them must be in $\overline{W}(\phi)$, a contradiction. Thus, there is at most one such orbit mapping to each end of S, and thus S is good. ☐

4. *Perturbation lemmas*

This section collects results on the robustness of certain phenomena under C^1 perturbation which we will use in section 5 to prove theorem A. Although the results in this section will be applied to flows satisfying conditions (3.1), we shall also apply them in §7 without assuming (3.1)

We begin with a digression, to clarify our topology on perturbations. The *strong C^r topology* is easiest to define when M is an open subset of the plane. A vectorfield X can then be regarded as a map from $\mathbb{R}^2$ to $\mathbb{R}^2$, and a basis for the strong C^r neighborhoods of X is given by the sets N(X, ε, r), as ε ranges over continuous strictly positive functions $\varepsilon : \mathbb{R}^2 \to \mathbb{R}$, where

$N(X, \varepsilon, r) = \{Y \mid \|D^i X(p) - D^i Y(p)\| < \varepsilon(p)$ for all p ε M and all i = 0, ..., r}.

Here, $D^0 X = X$ and $D^i X(p)$ is the ith derivative of X regarded as an i-linear $\mathbb{R}^2$-valued map, while $\| \cdot \|$ has its usual meaning as vector or (generalized) operator norm. Given a flow φ, we take N(φ, ε, r) as the set of flows ψ with $\dot{\psi} \in N(\dot{\phi}, \varepsilon, r)$. When M is not a subset of $\mathbb{R}^2$, we fix a locally finite cover by coordinate neighborhoods, use ε(p) as a positive number associated with each point, and calculate $\|D^i X(p) - D^i Y(p)\|$ in each local coordinate system. The topology so generated is independent of the choice of coordinate systems and metric on $\mathbb{R}^2$; it is not first countable (hence not metrizable), but has the Baire property [Pe3]. Since the function ε(p) can in general tend to zero as p tends "to infinity", the behavior of perturbations of X(p) is tightly controlled as p goes "to infinity".

This contrasts with the weaker *compact-open C^r topology*, in which a basis for the neighborhoods of X is given by sets

$N_0(X, \varepsilon, K, r) = \{Y \mid \|D^i X(p) - D^i Y(p)\| < \varepsilon$ for all p ε K and i = 0, ..., r}

where ε > 0 and K ranges over compact sets. Note that in this case the distinction between ε > 0 a number and a function is immaterial, since every such function has a positive lower bound on K. A compact-open neighborhood puts no control on behavior off a given compact set, but every compact-open neighborhood of X is open in the strong topology. Furthermore, we can define a strong neighborhood of X contained in the intersection of any countable family of compact-open neighborhoods, $N_0(X, \varepsilon_j, K_j, r)$ j = 1, ...,

provided they are *concentrated* on sets K_i which form a locally finite family
in M .

We begin with some simple adaptations of the compact theory, which we
formulate using the compact-open C^1 topology (so that we can later apply the
observation above), and state without proof. In what follows, $O_\phi[x,y]$
denotes the closed ϕ-orbit segment from x to y . The first two lemmas are
an application of the implicit function theorem.

(4.1) Lemma:

   Suppose $S_\pm$ are compact transverse sections to ϕ , and $p_\pm$ are inter-
ior points of $S_\pm$ with $p_+ \varepsilon O_+(p_-,\phi)$, There exists a compact-open C^1
neighborhood U of ϕ , concentrated on a given compact neighborhood of
$S_+ \cup S_- \cup O_\phi[p_-, p_+]$, and points $p_\pm(\psi)$ interior to $S_\pm$, varying contin-
uously with $\psi \varepsilon U,$ such that

      (i) $p_\pm(\psi) = p_\pm$

     (ii) $S_\pm$ are transverse sections for each $\psi \varepsilon U$

    (iii) $p_\pm(\psi) \varepsilon O_+(p_-(\psi),\psi)$

    (iv)   the arc $O_\psi[p_-(\psi), p_+(\psi)]$ varies C^1-continuously with $\psi \varepsilon U.$

(4.2) Lemma:

   Suppose $S_\pm$ are disjoint transverse sections to ϕ , such that:

      (i) S_+ is compact

     (ii)   the forward semi-orbit of each point x of S_+

             intersects S_- , and the first intersection

             $P_\phi(x)$ is interior to S_- .

Then the Poincaré map $P_\phi : S_+ \to int\ S_-$ is as smooth as ϕ, and there
exists a compact-open C^1-neighborhood U of ϕ , concentrated on a neighbor-
hood of the union of orbit segments $O_\phi[x, P_\phi(x)], x \varepsilon S_+,$ such that
$P_\psi : S_+ \to int\ S_-$ is well-defined and varies C^1-continuously with $\psi \varepsilon U.$

   The next lemma is the standard local stability of hyperbolic sinks and
sources:

(4.3) Lemma:

   Suppose γ is a fixed or periodic hyperbolic sink (resp. source) for
ϕ , and pick K a compact neighborhood of γ contained in its basin of
attraction $W^+(\gamma)$ (resp. region of repulsion $W^-(\gamma)$) with boundary ∂K
transverse to the flow. There exists a compact-open C^1 neighborhood U of ϕ ,
concentrated on K , and points or embedded circles $\gamma(\psi) \subset K$, varying (C^1)-
continuously with $\psi \varepsilon U,$ such that

      (i) $\gamma(\phi) = \gamma$

     (ii) $\gamma(\psi)$ is a fixed or periodic hyperbolic sink (source) for ψ

    (iii) $K \subset W^-(\gamma(\psi))$ $(K \subset W^+(\gamma(\psi)))$.

The corresponding fact for saddles uses the fact that compact parts of stable and unstable separatrices to fixed saddles vary C^r-continuously with the dynamical system [HP]. For x a fixed saddle and $y \in W^{\pm}(x)$, denote by $\sum(x, y)$ the open segment of separatrix joining x to y.

(4.4) *Lemma:*

Suppose x is a fixed hyperbolic saddle for ϕ and $y \in W^{\pm}(x)$. There exists a compact-open C^1 neighborhood U of ϕ concentrated on an arbitrarily given compact neighborhood of clos $\sum(x, y)$ and points $x(\psi)$, $y(\psi)$ varying continuously with $\psi \in U$ such that

> *(i) $x(\phi) = x$, $y(\phi) = y$*
>
> *(ii) $x(\psi)$ is a fixed hyperbolic saddle for ψ*
>
> *(iii) $y(\psi) \in W^{\pm}(x(\psi), \psi)$.*
>
> *(iv) the arc $\sum(x(\psi), y(\psi))$ varies C^1-continuously with $\psi \in U$.*

The phenomenon of interest to us which cannot be handled by standard compact arguments is escape to infinity. For a single orbit, the persistence of escape to infinity was handled by Collins [C], using Conley's notion of a Ważewski set [Co]. A related problem, the persistence of a dense set of separatrices, was handled in several cases by Peixoto-Pugh [PePu], Takens-White [TW], and Krych [Kr 3,4]. We present here a systematic argument incorporating features of several of these treatments.

Recall that a *flowbox* for a flow ϕ on a surface M is a closed quadrilateral $F \subset M$ containing no restpoints of ϕ, with two (opposite) edges $S_{\pm}$ transverse to ϕ and the other two edges ϕ-orbit segments, each joining an endpoint of S_+ to an endpoint of S_-. We call S_+ the *entrance set* and S_- the *exit set* of ϕ. It will be useful in the sequel to also consider *flow tubes*, defined as closed annuli containing no restpoints or periodic orbits such that the flow is transverse to their boundaries. The boundary of a flow tube consists of two circles, with the flow pointing in on one circle, the *entrance set* S_+, and out on the other circle, the *exit set* S_-. Thus, we can define a *generalized flow box* for ϕ as the region F formed by a union of orbit segments from an entrance transversal S_+ to an exit transversal S_- : F is a box or tube as S_+ (hence S_-) is an interval or circle. In what follows, any fussing about "edges" of F can be ignored when F is a tube.

To isolate a given positive or negative semi-orbit from others, and control its behavior under perturbation, we construct a positive (resp. negative) *tower*, defined as a finite or infinite sequence of generalized flowboxes $T = \{F_1, F_2, \ldots\}$ (resp. $T = \{F_{-1}, F_{-2}, \ldots\}$) satisfying, for each

$i = \pm 1, \ldots,$

(i) $F_i \cap F_j = \emptyset$ unless $|i - j| \leq 1$

(ii) $F_i \cap F_{i+1} = S_+(i + 1)$ (resp. $= S_-(i)$)

(iii) $S_+(i + 1) \subset$ int $S_-(i)$ (resp. $S_-(i - 1) \subset$ int $S_+(i)$)

(iv)   T forms a locally finite family in M.

The *floors* of the tower are the transverse edges $S_\pm(i)$ and its *height* is the number h of flowboxes (finite or infinite). A positive semi orbit can enter a positive tower T only via the bottom floor $S_+(1)$. An orbit can leave T via some set $S_-(i) \setminus S_+(i + 1)$, $i < h$, or else it crosses all flows of T before leaving T . Given a floor S of the positive tower T , we denote by *W(S, T, φ)* the set of all points $x \varepsilon S$ which cross all subsequent floors of T before leaving T . We adopt the analogous definitions when T is a negative tower. Note that for any tower T and any floor S , the set W(S, T, φ) is a nonempty closed interval (possibly a point), or circle and when T has infinite height, every semi orbit starting from W(S, T, φ) escapes to infinity inside T .

Note that if T is a tower for φ , it need not be a tower for flows ψ near φ , since the edges of the flowboxes need not be ψ-orbits. Nevertheless, if $\overset{\bullet}{\psi}$ is near $\overset{\bullet}{\phi}$ at points in T , we can still define the set *W(S, T, ψ)* as the set of points in S whose ψ-semi orbit crosses all subsequent floors of T in succession before leaving T . The following lemma is a persistence theorem for W(S, T, φ) . In (ii), $|J|$ denotes the length of the interval J . The result is formulated for positive towers; as usual, the reader can formulate the negative version.

(4.5) *Proposition:*

Suppose $T = \{F_1, \ldots\}$ is a positive tower for φ , and S is a floor of T .

(i)   *There exists a strong c^0-neighborhood U of φ (actually an intersection of compact-open neighborhoods concentrated on any neighborhood of the flowboxes of T) such that W(S, T, ψ) ≠ ∅ for ψ ε U.*

(ii)   *Given δ > 0 there exists a strong c^1-neighborhood U of φ (an intersection of compact-open c^1 neighborhoods concentrated on any neighborhood of the flowboxes of T) such that for every ψ ε U*
$(1 - δ)|W(S, T, φ)| \leq |W(S, T, ψ)| \leq (1 + δ)|W(S, T, φ)|$.

Proof:

Note that W(S, T, φ) is a nested intersection of intervals and/or circles $W_i(S, T, φ) \subset S$, defined as the set of points whose semi-orbit

crosses at least i successive floors in T.   The set in which these semi
orbits cross the ith floor S_i is an interval or circle $J_i \subset S_i$, and the
subset of J_i corresponding to $W_{i+1}(S, T, \phi)$ is the preimage by the
Poincaré map of S_{i+1} . Note that this set is interior to J_i , by condition
(iii) of the definition of tower.

For definiteness, denote by P_ϕ^{ij} $(i > j)$ the inverse Poincaré map of ϕ ,
from S_i to S_j . Thus,

$$P_\phi^{ij} = P_\phi^{i\ i-1} \circ P_\phi^{i-1\ i-2} \circ \ldots \circ P_\phi^{j+1\ j} \ .$$

Now, by (4.2) there are C^0 estimates concentrated on any neighborhood of
$F_i \cup \ldots \cup F_j$ which guarantee that ψ near ϕ defines a corresponding
Poincaré map $P^{ij} : S_i \to S_j$ which is C^0 near P_ϕ^{ij} , and C^1 near if ψ
is C^1-near ϕ on these sets. In particular, it is easy to see that by
estimates on the first i flowboxes we can insure that each Poincaré map
$P_\psi^{j+1,j}$ maps S_{j+1} into the interior of S_j , for all $j \leq i$. This guarantees in particular that

$$J_i(\psi) = P_\psi^{10} \circ P_\psi^{21} \circ \ldots \circ P_\psi^{i\ i-1}(S_i)$$

is a nonempty compact interval or circle interior to $J_{i-1}(\psi)$, and hence the
finite intersection property gives us conclusion (i).

To prove conclusion (ii), we note that for any monotone C^1 map f between intervals or circles the length of the image of a subinterval I is

$$\left| f(I) \right| = \int_I \left| f'(x) \right| dx \ .$$

Thus, if two maps $f, g : J \to J'$ satisfy an estimate of the form

(a) $(1 - \alpha)\left| f'(p) \right| \leq \left| g'(p) \right| \leq (1 + \alpha)\left| f'(p) \right| \ \forall \ p \ \varepsilon \ J$

then any interval $I \subset J$ satisfies

(b) $(1 - \alpha)\left| f(I) \right| \leq \left| g(I) \right| \leq (1 + \alpha)\left| f(I) \right| .$

Lemma (4.2) tells us that for any $\alpha < 1$ we can obtain (a) for $f = P_\phi^{i\ i-1}$,
$g = P_\psi^{i\ i-1}$ by controlling the C^1 distance between ϕ and ψ near F_i .

Thus, given $\delta > 0$, pick $\alpha_i > 0$ such that

$$1 - \delta < \prod_{i=1}^{h} (1 - \alpha_i)$$

$$1 + \delta > \prod_{i=1}^{\infty} (1 + \alpha_i)$$

and then make C^1 estimates on $\psi|F_i$ which insure that (a) (hence (b)) holds
for each i with $\alpha = \alpha_i$, $f = P_\phi^{i\ i-1}$, $g = P_\psi^{i\ i-1}$. By induction, we obtain
the analogue of conclusion (ii) for each set $W_i(S, T, \phi)$ and $W_i(S, T, \psi)$,
$i \leq h$, and hence (ii). $\square$

We note the following

(4.6) *Corollary: If a semi orbit $0_+(x, \phi)$ escapes to infinity, there
exists an infinite positive tower . T with $x \varepsilon S$, such that
$W(S_1, T, \phi) = \{x\}$, and hence for any flow ψ whose restriction to T is
sufficiently C^1-near $\phi|T$, there exists a unique ψ semi orbit which es-
capes to infinity inside T .*

To handle many semi orbits at once, we must combine many towers into a
positive (resp. negative) *palace*, defined as a locally finite collection P
of closed (generalized) flowboxes together with a partial ordering $F < \tilde{F}$
satisfying:

>   (i) $F < \tilde{F}$ iff there exists a positive (resp. negative) tower
>   of finite height starting at F and ending at $\tilde{F}$
>
>   (ii) $F \cap \tilde{F} \neq \emptyset$, $F \neq \tilde{F}$ implies F and $\tilde{F}$ are < -related
>
>   (iii) there is a unique $F_0 \varepsilon P$ (the <u>base</u> of the palace) such
>   that $F_0 < F$ for all $F \varepsilon P$.

We sketch a possible palace in fig. 4.1: note that a given transversal can be
the bottom floor of (even uncountably) many towers in P .

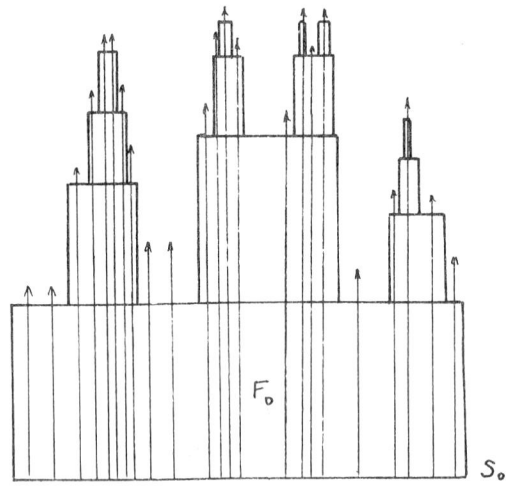

Figure 4.1

Given a floor S in P and a tower T in P built on S , define
$W(S, T, \phi)$ (and, for ψ near ϕ , $W(S, T, \psi)$) as before. Given a floor S ,
we define

$$W(S, P, \phi) = \bigcup W(S, T, \phi)$$

where the union is taken over towers T in P of maximal height. Note that
$W(S, P, \phi) \neq \emptyset$ for each S , but it need not be an interval or circle.

The following is an extension of (4.5) to palaces. Lebesgue measure is denoted meas . Note that a different Riemannian metric gives rise to an equivalent measure, so that the truth of (4.7) is metric-independent.

(4.7) *Proposition*:

Given a palace P *for* ϕ , *there exists a strong* C^1-*neighborhood* U *of* ϕ *(an intersection of compact-open* C^1 *neighborhoods concentrated on the flowboxes of* P*) such that for every floor* S *of* P *and every* $\psi \in U$

(i) $W(S, P, \phi)$ *and* $W(S, P, \psi)$ *are homeomorphic, preserving towers,*

and furthermore given $\delta > 0$, U *can be chosen so that for every* S *and* $\psi \in U$,

(ii) $(1 - \delta)$ *meas* $[W(S, P, \phi)] \leq$ *meas* $[W(S, P, \psi)]$
$$\leq (1 + \delta) \text{ meas } [W(S, P, \phi)].$$

Proof:

This is a straightforward adaptation of (4.5). Note that, given a flowbox $F \in P$, we can assign it a "height" in P : the height of the tower from the base F_0 to F. Given a finite height h , there exist only finitely many flowboxes $F \in P$ with height $\leq$ h. This allows us to carry out the box-by-box estimates from the proof of (4.5) simultaneously for all towers. In particular, we can insure that for each tower T in P, $W(S, T, \phi)$ and $W(S, T, \psi)$ are both either a point (length 0) or an interval with interior (length > 0). This lets us construct the homeomorphism of (i). To prove (ii), note that we can express any $W(S, P, \psi)$ as a nested intersection $\bigcap_{i=1}^{h} W_i(S, P, \psi)$ where $W_i(S, P, \psi)$ is the union, over all towers T built on S, of $W_j(S, T, \psi)$, j = min (i, height T) . But $W_i(S, P, \psi)$ for i fixed is a finite union of intervals, each varying continuously with ψ . We can insure the estimate on the measure of each $W_i(S, T, \psi)$, touching only finitely many flowboxes in the process. Then (ii) appears in the limit. □

(4.8) *Remark*:

(4.7) (ii) allows us to complete the argument for example (2.7). We simply build positive palaces on S_{-1} , S_0 and negative palaces on S_0, S_+ for which the sets $W(S_t, P, \phi)$ are $W^{\pm}(\phi) \cap S_t$. Our construction gives meas $W(S_t, P, \phi) = 3/4$. so that for ψ near ϕ estimate (ii) above gives meas $W(S_t, P, \psi) > 1/2$ and the rest of (2.7) follows.

5. *Proof of Theorem A*

We prove the structural stability of a flow ϕ satisfying (3.1) in two steps. First, we use the results of sections 3 and 4 to construct neighborhoods of the sinks, sources, fixed saddles, and stable and unstable separatrices which constitute a kind of fine filtration for the "separatrix structure" of ϕ. Using this and the results of section 4 it is easy to see that this separatrix structure persists under perturbations. W. Kaplan [Ka], F. Pluvinage [Pℓu], L. Markus [M1] and D. Neumann [Ne2, NeO] have shown that two-dimensional flows with equivalent separatrix structures are topologically equivalent. We use the method of Neumann to construct an equivalence homeomorphism h between ϕ and its perturbation ψ. An examination of this construction shows that by making ψ strongly C^1 near ϕ we insure that h lies in a given compact-open neighborhood of the identity.

The first step of our proof closely resembles the global part of the standard proof of structural stability for Morse-Smale flows on closed surfaces. In that case, local stability near the periodic orbits (roughly our lemmas (4.3-4)) combines with a kind of filtration to give global stability. Since every orbit in this case has an α- and ω-limit set in $Per(\phi)$, the construction of a global equivalence homeomorphism proceeds from local ones near $Per(\phi)$ by consideration of where orbits tend in both time directions.

In our case, in addition to α- and ω-limits at $Per(\phi)$, we can witness several varieties of escape to infinity. We have noted some analogy with the compact case by indicating when a semi orbit is regarded as tending toward a saddle at infinity. Let us extend the analogy by saying a positive (negative) semi orbit $O_{\pm}(x)$ tends to a *sink (source) at infinity* if some neighborhood of x escapes to infinity uniformly in forward (backward) time. Prop. (3.9) tells us that under hypotheses (3.1) every positive semi orbit not in clos $W^+(\phi)$ tends toward a sink, either in $Per(\phi)$ or at infinity. Every positive semi orbit in $W^+(\phi)$ tends toward a saddle (fixed or at infinity). Note, however, that semi orbits in $[\text{clos } W^+(\phi)] \setminus W^+(\phi)$ do neither.

Our first step consists of constructing two pairs of nested closed submanifolds with boundary transverse to ϕ, $M_1^+ \subset M_2^+$ and $M_1^- \subset M_2^-$, such that $M_i^-(M_i^+)$ is forward (backward) invariant under the flow, $M_1^-(M_1^+)$ is an isolating neighborhood for the sinks (sources), $M_2^-(M_2^+)$ isolates the sinks (sources)

44

together with clos $W^-(\phi)$ (clos $W^+(\phi)$), and $M_2^+ \cap M_2^-$ isolates the fixed
saddles; M_1^+ and M_1^- are disjoint, while any orbit leaving M_2^+ enters M_2^- .
These isolating sets similarly isolate the corresponding sets for any per-
turbation ψ of ϕ . The sets $M_1^{\pm}$ are bounded by *global sections* $\sum_{\pm}$ which
are specially chosen unions of cross sections to the semi-canonical regions.
(These $\sum_{\pm}$ are technically not "transverse sections" or "cross sections" because they have several components.) When all saddles are at infinity, the
sets $M_2^{\mp}$ are disjoint, and are constructed by first building a negative
(resp. positive) *grand palace* P_+ (resp. P_-) with base $\sum_+$ $(\sum_-)$ and then
"smoothing" the edges. The adjective "grand" here indicates that $P_{\pm}$ differs
technically from our earlier definition in that the base is not connected.
(Thus, a "grand palace" is a union of "palaces" such that each point has a
neighborhood intersecting at most one palace, and no orbit leaving one palace
can enter another.) When fixed saddles are present, we truncate the towers of
$P_{\mp}$ containing a stable or unstable separatrix of each fixed saddle γ_i , and
surmount these finite towers with quadrilaterals Q_i containing γ_i . The
sets $M_2^{\pm}$ are then formed by smoothing the edges of $P_{\pm} \cup Q$, where
$Q = \cup_i Q_i$.

The construction of $\sum_{\pm}$ and $P_{\pm}$ is easiest to formulate when all sinks,
sources and saddles are at infinity. Thus, given a general flow ϕ on M
satisfying (3.1), we first consider a new flow $\widetilde{\phi}$ on a new open surface $\widetilde{M}$
obtained by removing $\Omega(\phi)$ from M . It is clear that

      (i) $\widetilde{\phi}$ is *completely unstable*: $\Omega(\widetilde{\phi}) = \emptyset$

      (ii) clos $W^{\pm}(\widetilde{\phi}) = \widetilde{M} \cap$ clos $W^{\pm}(\phi)$.

The reader is cautioned that structural stability of $\widetilde{\phi}$ does not automatically
imply structural stability for ϕ , since the strong topology on $\widetilde{M}$ controls
perturbations much more strongly near $\Omega(\phi)$ than does the strong topology on
M . However, we shall use $\widetilde{\phi}$ only as an aid to constructing $\sum_{\pm}$ and $P_{\pm}$
by assuming $\Omega(\phi) = \emptyset$ in (5.1-5.3). We will return to ϕ and M (and,
where necessary, introduce the quadrilateral set Q) before considering
perturbations.

The sets $\sum_{\pm}$ are almost constructed, separately, in (3.15), but to re-
late them to each other we need the following technical result.

(5.1) *Lemma*:

    Suppose $\Omega(\phi) = \emptyset$ *and* ϕ *satisfies* (3.1). *Then there exist two families
of compact transverse sections* $\{s_i^{\pm}\}$ *satisfying*:

      (i) $\{s_i^{\pm}\}$ *form a locally finite family, and their endpoints
         belong neither to clos $W^+(\phi)$ nor to clos $W^-(\phi)$.*

      (ii) *No orbit crosses $s_i^{\pm}$ twice.*

(iii) *for each* i,
$$S_i^+ \cap W^-(\phi) = \emptyset = S_i^- \cap W^+(\phi)$$
and
$$S_i^{\pm} \cap W^{\pm}(\phi) \neq \emptyset$$

(iv) *clos* $W^{\pm}(\phi) \subset \bigcup_i Sat\ (S_i^{\pm})$

(v) $Sat\ (S_i^+) \cap Sat\ (S_j^+) = \emptyset$ *for* $i \neq j$,

$Sat\ (S_i^-) \cap Sat\ (S_j^-) = \emptyset$

(vi) $Sat_-\ (S_i^+) \cap Sat_+\ (S_j^-) = \emptyset$ *for all* i, j.

Proof:

It is easy to pick transversals $S_i^{\pm}$ through points of $W^{\pm}(\phi)$ so that (i)-(iv) hold. We must modify these without affecting (i)-(iv) so as to obtain (v) and (vi). We do this inductively for $n = 1, 2, \ldots$.

Suppose we have $\{S_i^{\pm}\}$ so that (i)-(iv) hold and

(v, n) $i < n$, $j \neq i$ imply
$$Sat\ (S_i^+) \cap Sat\ (S_j^+) = \emptyset = Sat\ (S_i^-) \cap Sat\ (S_j^-)$$

(vi, n) $i < n$, $j < n$ imply
$$Sat_-\ (S_i^+) \cap Sat_+\ (S_j^-) = \emptyset \ .$$

We can easily shrink all $S_j^{\pm}$, $j > n$ so that (v, n+1) holds and (i)-(iv) are unaffected. This also guarantees all conditions of (vi, n+1) except

(vi, n+1, 1) $Sat_-\ (S_n^+) \cap Sat_+\ (S_j^-) = \emptyset$ $j = 1. \ldots, n$

(vi, n+1, 2) $Sat_-\ (S_i^+) \cap Sat_+\ (S_n^-) = \emptyset$ $i = 1, \ldots, n$.

The first of these is a finite set of conditions on $Sat_-\ (S_n^+)$. Note that by (3.8) or (3.9) S_n^+ escapes to infinity uniformly in backward time. Thus, for each S_j^-, $j = 1, \ldots, n$, there exists a time T_j so that

$\phi(t, S_n^+) \cap S_j^- = \emptyset$ for $t < T_j$. Let $T = \min\limits_{i=1,\ldots,n} T_j$ and replace S_n^+ with $\phi(T-1, S_n^+)$. Then $Sat\ (S_n^+)$ (hence (i)-(iv)) is unaffected, but (vi, n+1, 1) now holds. Then, a similar argument for S_n^- allows us to replace it with some set $\phi(T'+1, S_n^-)$ for which (vi, n+1, 2) holds.

Note that this inductive process does not alter $S_n^{\pm}$ after the nth step, so that it ultimately results in a set of transversals as required. □

Note that even though the sections S_i^+ form a locally finite family, their saturations do not. However, these saturations are locally finite away from $clos\ W^-(\phi)$.

(5.2) *Lemma:*

If $\Omega(\phi) = \emptyset$ *and* ϕ *satisfies* (3.1), *given the families of transverse sections* $\{S_i^{\pm}\}$ *as in* (5.1) *there exist closed sets* $\sum_{\pm}$ *satisfying*

(i) $\sum_+ \cap clos\ W^-(\phi) = \emptyset = \sum_- \cap clos\ W^+(\phi)$

(ii) *the intersection of* $\sum_{-}$ $(\sum_{+})$ *with any positive*

   (negative) semi-canonical region A *for* ϕ *is*

   a cross section for A.

(iii) *for every* i, $s_i^{\pm} \subset Sat_{\pm}$ $(\sum_{\pm})$.

(iv) *if* $x \varepsilon \sum_{+}$ *and* $y \varepsilon O(x) \cap \sum_{-}$,

   then $y \varepsilon O_{+}(x)$.

Proof:

We construct $\sum_{-}$ and leave the dual construction of $\sum_{+}$ to the reader. Using (3.15), we can construct $\sum_{-}'$ satisfying (i) and (ii) by taking a union of good cross-sections for positive semi-canonical regions. Note that $\sum_{-}'$ may not be closed, as endpoints of good cross-sections might belong to $W^{-}(\phi)$. Consider the sets $\sum_{i}' = \sum_{-}' \cap sat$ (S_i^{-}), which are disjoint and locally finite on $\sum_{-}'$ (see the comment preceding the statement of (5.2)). By (3.9), there exists T_i such that $\phi(t, \sum_{i}') \cap S_i^{-} = \emptyset$ for $t > T_i$. By local finiteness, we can find a C^{∞} function $T(x) \geq 0$ on $\sum_{-}'$ such that $T(x) > T_i$ for $x \varepsilon \sum_{i}'$. Then replace $\sum_{-}'$ with

$$\sum_{-}'' = \{\phi(T(x), x) \mid x \varepsilon \sum_{-}'\}$$

which satisfies (iii).

Having constructed $\sum_{-}''$ and its dual $\sum_{+}''$, it remains to modify them so that neither has endpoints in the wrong set $W^{\pm}(\phi)$ and (iv) holds. To this end, express each as a nested union of compact sets

$$\sum_{\pm}'' = \bigcup_{i=1}^{\infty} \sum_{\pm}(i)$$

and apply a process similar to that in the preceding proof: replace $\sum_{-}$ (i) $(\sum_{+}$ (i)$)$ with a forward (backward) image so that (iv) holds on $\sum_{\pm}$ (i) — that is,

$$Sat_{-} (\sum_{-} (i)) \cap Sat_{+} (\sum_{+} (i)) = \emptyset$$

By a similar process, we can insure that at the "endpoints" of $\sum_{\pm}$ where there is some danger of running into $W^{\mp}(\phi)$, the section escapes to infinity. This insures that $\sum_{\pm}$ are closed, and satisfy (i)-(iv). $\square$

We now construct the grand palaces $P_{\pm}$, still under the assumption that ϕ is completely unstable:

(5.3) *Proposition:*

Given a flow ϕ *with* $\Omega(\phi) = \emptyset$ *and satisfying (3.1), and given* $\{s_i^{\pm}\}$ *as in (5.1) and* $\sum_{\pm}$ *as in (5.2), there exist a positive and negative grand palace* $P_{\pm}$ *with base* $\sum_{\pm}$ *such that*

   (i) $P_{+} \cap P_{-} = \emptyset$

(ii) $W(\sum_{\pm}, P_{\pm}, \phi) = \sum_{\pm}^{*} \cup [\sum_{\pm} \cap clos\ w^{\pm}(\phi)]$, *where* $\sum_{\pm}^{*}$

  consists of circles escaping uniformly to infinity

  in both time directions.

Proof:

Given $\sum_{\pm}$, define functions

$$\tau_{\pm} : [\sum_{\pm} \setminus clos\ w^{\pm}(\phi)] \to \mathbb{R}^{+}$$

by

$$\phi(\pm\ \tau_{\pm}(x),\ x) \in \sum_{\mp},\quad x \in \sum_{\pm} \setminus clos\ w^{\pm}(\phi)\ .$$

These functions are well-defined and C^{1}, and their limit at points of
$\sum_{\pm} \cap clos\ w^{\pm}(\phi)$ is infinite. In fact, if $x \in \sum_{+}$ and $y \in O(x) \cap \sum_{-}$ then
$\tau_{+}(x) = \tau_{-}(y)$ equals the time length of the orbit segment $O[x,y]$. For con-
venience, we define $\tau_{\pm}$ as a continuous extended-real-valued function on all
of $\sum_{\pm}$ by setting $\tau_{\pm} = +\infty$ at $\sum_{\pm} \cap clos\ w^{\pm}(\phi)$.

We first eliminate the circles $C_{\pm}$ in $\sum_{\pm}^{*}$ by noting that they come in
pairs, and $\tau_{\pm}$ has a positive minimum m on each. For these, we define a
single flow tube in each palace, whose exit set is $\phi(\pm\frac{m}{3},\ C_{\pm})$.

By Sard's theorem, we can find regular values n_{i}, $i = 1,\ \dots$ of $\tau_{\pm}$
such that

$$|n_{i} - i| < 1/10\ .$$

Now, define a nested sequence of closed subsets

$$\sum_{\pm}\ (i) \subset \sum_{\pm}$$

by

$$\sum_{\pm}\ (i) = \{x \in \sum_{\pm}\ |\ \tau_{\pm}(x)\ \geq n_{i}\}\ ,\ i \geq 1$$

(The reader is cautioned that these sets behave like the complements of the
sets denoted the same way in the previous proof.) Note that
$\sum_{\pm}\ (i+1) \subset int\ \sum_{\pm}\ (i)$, each set is the closure of its interior (by regularity
of n_{i}), each is a locally finite family of closed intervals, and a given
point of $\sum_{\pm} \setminus clos\ w^{+}(\phi)$ has a neighborhood intersecting only finitely many
$\sum_{\pm}\ (i)$.

Now, define the rest of $P_{\pm}$ so that the nth floor of $P_{\pm}$ is

$$S_{\pm}(n) = \phi(\pm\ n/3,\ \sum_{\pm}(n))\ .$$

Thus, a given orbit segment passing from $x_{+} \in \sum_{+}(n)$ to $x_{-} \in \sum_{-}(n)$ leaves
P_{+} before the end of the first third of $O[x_{+},\ x_{-}]$ and enters P_{-} in its
last third. Thus, $P_{\pm}$ are disjoint by construction, and $W(\sum_{\pm},\ P_{\pm},\ \phi)$ is
the set of points where $\tau_{\pm} = \infty$, in other words $\sum_{\pm} \cap clos\ w^{\pm}(\phi)$, as required. Local finiteness of $P_{\pm}$ is insured by the fact that the only possible

accumulation of flowboxes of, say P_- is on clos $W^+(\phi)$, from which P_- is separated by P_+. ☐

We now return to the general case in which $\Omega(\phi)$ may be nonempty.

(5.4) <u>Lemma</u> (Fig. 5.1):

Suppose ϕ satisfies (3.1). There exist a closed neighborhood Q of the fixed saddles consisting of disjoint quadrilaterals Q_i and two disjoint grand palaces, P_- (resp. P_+) negative (resp. positive) with base $\textstyle\sum_+$ (resp. $\textstyle\sum_+$), such that:

(i)   Each Q_i contains a unique fixed saddle, γ_i, has edges transverse to the flow, and all orbits other than γ_i and its stable (resp. unstable) separatrices leave Q_i in forward (resp. backward) time;

(ii) The intersection of $\textstyle\sum_-$ ($\textstyle\sum_+$) with a positive (resp. negative) semi-canonical region A is a closed cross-section for A.

(iii) Each finite tower of $P_\pm$ with non-circular floors ends in a (different) quadrilateral Q_i.

(iv) Every orbit leaving $P_\pm \cup Q$ enters $P_\mp$; the points whose orbits fail to leave $P_+ \cup Q$ (resp. $P_- \cup Q$) in forward (resp. backward) time constitute the complement of the sources (resp. sinks) in clos $W^+(\phi)$ (resp. clos $W^-(\phi)$).

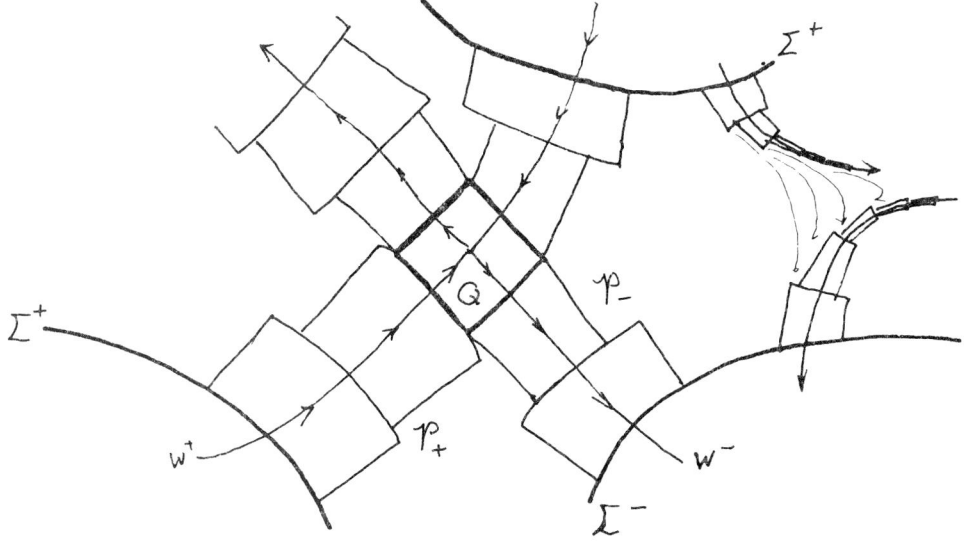

Figure 5.1

Proof:

We begin by applying (5.2)-(5.3) to the flow $\widetilde{\phi} = \phi\,|\,\widetilde{M}$, $\widetilde{M} = [M \setminus \Omega(\phi)]$. This gives us grand palaces $\widetilde{P}_\pm$ with floors $\widetilde{\Sigma}_\pm$, which may however fail to be locally finite in M at points of $\Omega(\phi)$.

Note, first, that if ϕ has a sink (resp. source) γ disjoint from clos $W^-(\phi)$ (resp. clos $W^+(\phi)$) then the corresponding componet of Σ_- (resp. Σ_+) is a circle around γ. We have no problem in this case. If a sink (source) γ has unstable (resp. stable) separatrices tending toward it, we pick a circle C about γ, contained in $W^-(\gamma)$, and replace any part of $\Sigma_{\pm}$ inside C with the intersection of its backward saturation with C. This can be smoothed at the ends, so that γ is not an accumulation point of $\Sigma_{\mp}$, and by picking C small, we still don't affect the conclusions of (5.2-3).

It is clear that we can pick $Q = \bigcup Q_i$ so that (i) holds, and Q is disjoint from all the circles C and sections $\Sigma_\pm$. We have the problem that the separatrices of the saddle γ_i will live in infinite towers of $\widetilde{P}_\pm$ accumulating at γ_i. However, by shrinking Q_i if necessary we can insure that each edge of Q_i is contained in the appropriate saturation of a single floor of $\widetilde{P}_\pm$. Then we can truncate the tower of this flow and build a flowbox connecting it to an edge of Q_i. Denote these modified grand palaces by $P_\pm$. The reader can easily verify that (i)-(iv) are now satisfied. ☐

The structure we want is essentially that of (5.4), but it is easier to handle perturbations if instead of towers and palaces, with edges which are ϕ-orbit segments, we "smooth out" the edges so that they are C^1 curves transverse to the flow. This is an essentially local (and locally finite) process. Note that no smoothing is necessary (since there are no edges) in a tower consisting of flow tubes. On the other hand, suppose a (positive) palace contains the flowbox F_0, with several successors $F_1, \ldots, F_n$. Thus the entrance sets $S_+(i)$, $i=1, \ldots, n$ of F_i are closed disjoint intervals in the interior of the exit set $S_-(0)$ of F_0. We wish to "shave" F_0 (and, ultimately, F_i $i=1 \ldots, n$) so that the new "vertical" edges are transverse to the flow. To this end, denote by $\tau(x)$ the time of passage across F_0: $\phi(x, \tau(x)) \varepsilon S_-(0)$ for $x \varepsilon S_+(0)$. Parametrize $S_+(0)$ and $S_-(0)$ by $[0,1]$ so that the same parameter value corresponds to the same orbit, and let $[a,b] \subset S_-(0)$ be the smallest interval containing all the sets $S_+(i)$, $i=1, \ldots, n$. Then $0 < a < b < 1$. Pick a C^∞ bump function θ on $\mathbb{R}$ so that $\theta^{-1}(1) = [a,b]$ and $\theta^{-1}(0) = (-\infty, 0] \cup [1, \infty)$. Then the set

$$F_0' = \{\phi(x,t) \mid x \in S_+(0), \ 0 \le t \le \theta(x) \tau(x)\}$$

has all edges transverse to the flow: there is the entrance set $S_+(0)$, and
a smooth curve composed of entrance sets $S_+(i)$, $i=1, \ldots, n$ for the next
flowboxes, and escape edges by which the flow transversally leaves the tower.
(see fig. 5.2)

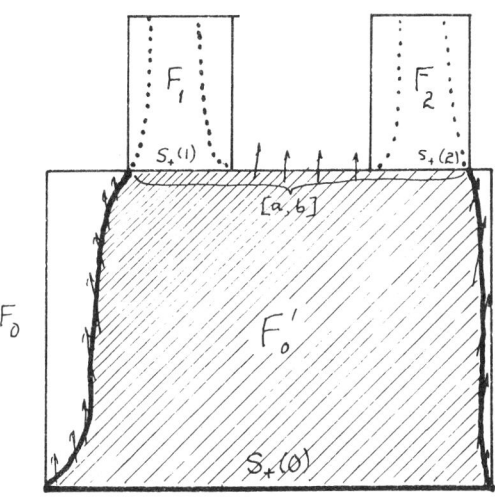

Figure 5.2

If we apply this process in succession to the flowboxes of a (grand) palace,
P, we obtain a structure $M(P) \subset P$ which has the same base $\sum$ as P, and
the top floor of each finite tower of P is an exit set of $M(P)$, while the
rest of the boundary of $M(P)$ consists of C^1 escape edges. If we define
the floors S of $M(P)$ and the sets $W(S, M(P), \phi)$ by analogy with those
of P, we see that the construction leaves

$$W(\textstyle\sum, \ M(P), \ \phi) = W(\textstyle\sum, \ P, \phi).$$

We can apply this process to our truncated palaces $P_\pm$. If a finite
tower ends in a quadrilateral Q_i, then we smooth the meeting with the top
floor in such a way that the escape edge is tangent to the exit (instead of
entrance) set of the quadrilateral Q_i. (see fig. 5.3). Note that each
quadrilateral Q_i joins two towers of P_+ (resp. P_-) into a "handle"
transverse to ϕ.

This construction gives us the structure described in the following

(5.5) *Corollary (Fig. 5.4):*

If ϕ satisfies (3.1), there exist closed 2-dimensional C^1 submani-
folds-with-boundary in M, $M_1^- \subset M_2^-$ and $M_1^+ \subset M_2^+$, such that:

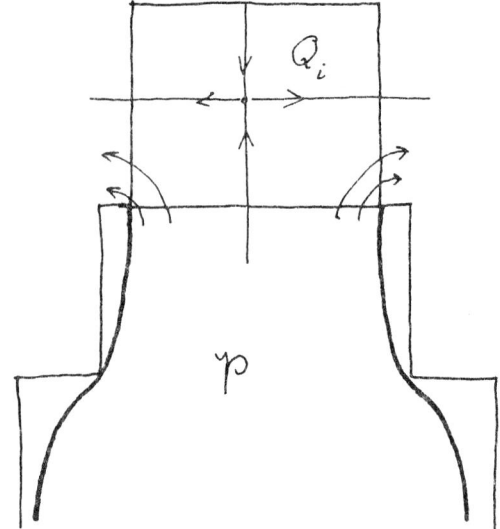

Figure 5.3

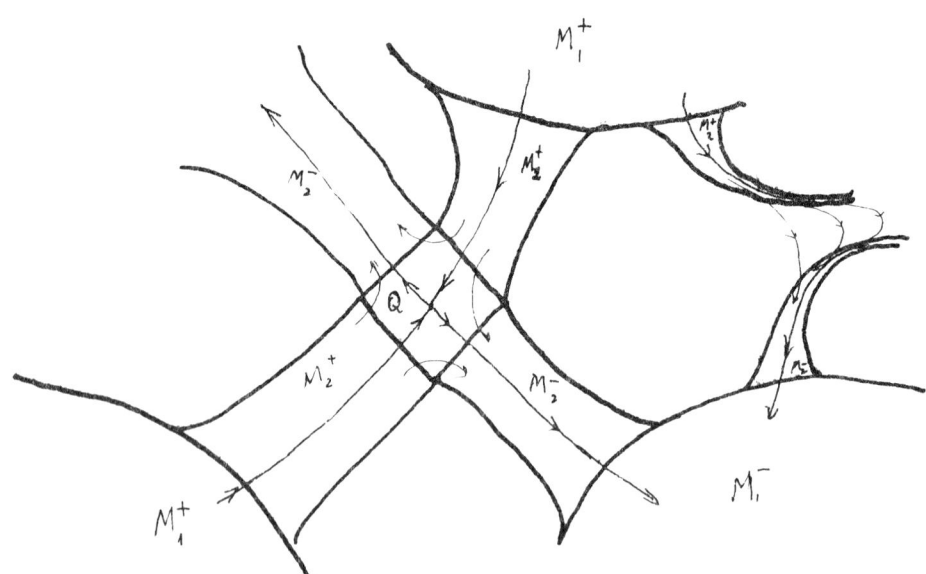

Figure 5.4

(i) M_i^- (M_i^+) is forward (backward) invariant under
     ϕ , for $i=1, 2$.

(ii) $\partial M_1^+ = \sum_\pm$ is a global section whose intersection
      with any positive (negative) semi-canonical region
      is a cross-section for it. In particular, every
      component of M_1^+ is either contained in $W^\pm(\sigma)$
      for some sink (source) $\sigma \in Per(\phi)$, or else
      escapes to infinity uniformly on compacta in posi-
      tive (resp. negative) time.

(iii) any orbit contained in M_1^- (M_1^+) is a sink (source)
       in $Per(\phi)$, and any other semi-orbit not tending
       toward a saddle (fixed or at infinity) enters
       M_1^- (M_1^+) in forward (backward) time.

(iv) ∂M_2^+ is transverse to $\dot\phi$. M_2^- contains $clos\ W^+(\phi)$
      together with the sinks (sources) in $Per(\phi)$, and
      no other complete orbits.

(v) The stable (unstable) separatrices of saddles at
     infinity do not enter M_2^- (M_2^+) .

(vi) $M_2^+ \cap M_2^- = Q$ is a union of quadrilaterals Q_i ,
      each isolating a unique fixed saddle γ_i .

From this structure, the proof of theorem A is relatively straightforward. We re-state the result in full:

(5.6) Theorem A:

Suppose ϕ is a C^1 flow on the surface M such that

(i)   there are no non-trivial minimal sets and
      no oscillating orbits

(ii) every orbit in $Per(\phi)$ is hyperbolic

(iii) $clos\ W^-(\phi) \cap clos\ W^+(\phi) \subset Per(\phi)$.

Then $\Omega(\phi) = Per(\phi)$ and given a compact-open neighborhood N of the identity, there exists a strong C^1 neighborhood U of ϕ such that every flow
$\psi \in U$ is topologically equivalent to ϕ via a homeomorphism $h \in N$.

Proof of theorem A:

Construct the sections $\sum_\pm$ grand palaces $P_\pm$, and quadrilaterals Q_i
as in (5.2-4), and smooth them into the sets $M_1^\pm$, $M_2^\pm$, $Q = \bigcup Q_i = M_2^+ \cap M_2^-$
as in (5.5). Let $R^\pm \subset int\ M_1^\mp$ be neighborhoods of the sinks and sources in
$Per(\phi)$, such that each component $R_i^\mp$ of R^- (R^+) is a compact neighborhood
of a single sink (source) σ_i , with the boundary of $R_i^\mp$ one or two circles

transverse to ϕ . Note that the smoothed palaces $M_2^\pm \setminus M_1^\pm \subset P_\pm \cup Q$ have a locally finite decomposition into quadrilaterals Q_i , flow tubes and "shaved" flow boxes. Each such generalized flowbox F has a well-defined height $h(F)$ relative to the base $\sum_\pm$.

We define a strong C^1 neighborhood U of ϕ by the following locally finite sets of conditions on $\psi \; \varepsilon \; U$:

    (i) ψ is transverse to $\partial M_i^\pm$ for i=1, 2 .

    (ii) ψ is transverse to the boundaries of $R^\pm$ and Q , and in each set $R_i^\mp$, Q_i ψ has a unique source (resp. sink, saddle) with $R_i^\pm \subset W^\mp(\sigma_i^\pm(\psi))$, and each separatrix of $\gamma_i(\psi)$ leaves Q_i (use 4.3-4)

  (iii) For each tower T of $P_\pm$, $W(\sum_\pm , T , \psi)$ is homeomorphic to $W(\sum_\pm , T , \phi)$ (use 4.7).

   (iv) Every ϕ-orbit segment from the escape set of a given (generalized, shaved) flow box F_+ of $M_2^+ \setminus M_1^+$ to the escape set of a given flow box F_- of $M_2^- \setminus M_1^-$ is C^1-near the ψ-orbit segment starting from the same point in F_+ , which enters M_2^- (use 4.1, since these segments are compact and note that the components of $M \setminus [P_+ \cup P_-]$ form a locally finite family).

   (v) Every ϕ-orbit segment $O_\phi[x,y]$ with $x \; \varepsilon \; S_\mp(F_\mp)$ and $y \; \varepsilon \; S_\pm(F'_\mp)$, where $F'_\mp$ is a successor of $F_\mp$ in the palace $P_\mp$, is C^1-near an orbit segment $O_\psi[x',y]$ with $x' \; \varepsilon \; S_\mp(F_\mp)$. (use 4.2)

The local finiteness of all these conditions implies that they do indeed define a strong C^1 neighborhood U of ϕ . Now, any separatrix of a saddle at infinity for ϕ is isolated by an infinite tower of $P_\pm$ (hence of $M(P_\pm)$), and (4.6) finds us a unique corresponding orbit for $\psi \; \varepsilon \; U$. That the ψ-orbits corresponding to the ϕ-separatrices of a saddle at infinity are ψ-separatrices follows from conditions (iv) and (v) above, since ψ-orbits near the alleged stable separatrix crossing floors of P_+ with large height enter the tower of P_- isolating the alleged unstable separatrix, at large height. Thus, $\operatorname{clos} W^\pm(\psi) \cap \sum_\pm = W(\sum_\pm , P_\pm , \psi)$ for all $\psi \; \varepsilon \; U$.

We see, then, that any $\psi \; \varepsilon \; U$ has the same separatrix structure as ϕ , in the sense of D. Neumann [Ne 2]. Under these circumstances Neumann shows how to construct an equivalence homeomorphism h between ϕ and ψ . We outline this construction, to see how it gives h ε N .

Neumann's basic idea is to decompose each canonical region A into quadrangles whose size is bounded on A by a constant $\sigma(A) > 0$ and goes to zero at the boundary of A. For two flows with identical canonical regions, this decomposition gives a correspondence between cells which is the identity on the boundary of A. In particular, $\sigma(A)$ measures the distance between h and the identity at points of A. There are countably many canonical regions altogether, so we can number them A_i, $i=1, \ldots$, and construct our decompositions so that $\sigma(A_i) < 1/i$. This insures that the equivalence h defined on canonical regions extends continuously to limit separatrices as the identity map.

Our constructions in (5.1-5) insure that given a metric on M, we can insure (by making ψ strongly C^1-near ϕ) that the separatrix set $\Omega(\psi) \cup \text{clos}[W^-(\psi) \cup W^+(\psi)]$ of ψ is homeomorphic to that of ϕ by a map h_1 uniformly (C^0) near the identity (since towers go to towers in $P_\pm$). We can therefore replace the C^1 flow ψ in our equivalence problem with a C^0 flow ψ_1 conjugate to ψ via h_1 and having separatrix set identical to that of ϕ.

Now, given the compact set K on which N is concentrated, we can control our cell-decomposition so that every cell intersecting K has an a priori uniformly bounded size. Then the equivalence h_2 between ϕ and ψ_1 is uniformly (C^0) near the identity on K, and h_1 is uniformly (C^0) near the identity on M, so that $h = h_1 \circ h_2$ is (C^0) uniformly near the identity on K, and thus belongs to N.

This proves that ϕ is C^1 globally structurally stable, as required. $\square$

6. *Transverse sections to planar flows*

Our attention in this and the next two sections will be confined to flows
in the plane, as we prove theorem B. The present section investigates the use
of transversals in determining the global behavior of certain orbits. The
next section presents a sequence of genericity results, including proofs of
(1.5) and theorem C. Finally section 8 presents theorem B and a few other
technical facts concerning necessary conditions for structural stability of
flows in the plane.

The transverse sections we consider will all be embedded compact intervals. Any such interval has two natural orderings, and we will adopt one of
these at will for notation purposes, without further comment. We will denote
segments of transversals in a manner similar to our earlier notation (which
we also continue to use) for orbit segments: if a, b ϵ S , where S is an
interval transverse to the flow, then

$$S[a,b] = \{x \; \epsilon \; S \mid a \leq x \leq b\}$$
$$S(a,b) = \{x \; \epsilon \; S \mid a < x < b\}$$

where "$\leq$" is one of the two orders on S . Note that no special significance is attached to the choice of order, so $S[a,b] = S[b,a]$.

Our basic tool is the following consequence of the Jordan curve theorem:

(6.1) *Lemma:*

Suppose ϕ *is a flow in* $\mathbb{R}^2$ *and* S *is an interval transverse to* ϕ *.*
If $a_i \; \epsilon \; S$, *i=1,2 are successive intersections of a single* ϕ-*orbit with* S ,
then $L = 0[a_1,a_2] \cup S[a_1,a_2]$ *separates* $\mathbb{R}^2$ *into two connected open sets,*
$A_{\pm}$, *where* $\phi(t,x) \; \epsilon \; A_{\pm}$ *for all* $x \; \epsilon \; A_{\pm}$ *and* $\pm t > 0$.

Proof:

L is a simple closed curve, so $\mathbb{R}^2 \setminus L$ has two components, $A_{\pm}$.
Since $0[a_1,a_2]$ is an orbit segment, ϕ-orbits can pass between these components via $S[a_1,a_2]$, but this only occurs in one direction. $\square$

(6.1a) *Remark:*

For any flow ϕ *in the plane, if* $0_+(x,\phi)$ *oscillates, then* $\alpha(x,\phi)$ *is*
bounded and nonempty.

This is because for any transverse section S through a point of $\omega(x,\phi)$,
$0_+(x,\phi)$ crosses S twice, say at x_1 and x_2 , and by (6.1) the plane is

separated by $S[x_1,x_2] \cup 0[x_1,x_2]$ into two invariant regions, $A_\pm$, which are positively (resp. negatively) invariant. A_+ contains all of $\omega(x_2,\phi) = \omega(x,\phi)$, hence is unbounded. This says $0_-(x_1,\phi) \subset A_-$ is bounded, and so $\alpha(x,\phi) = \alpha(x_1,\phi) \neq \emptyset$ is compact.

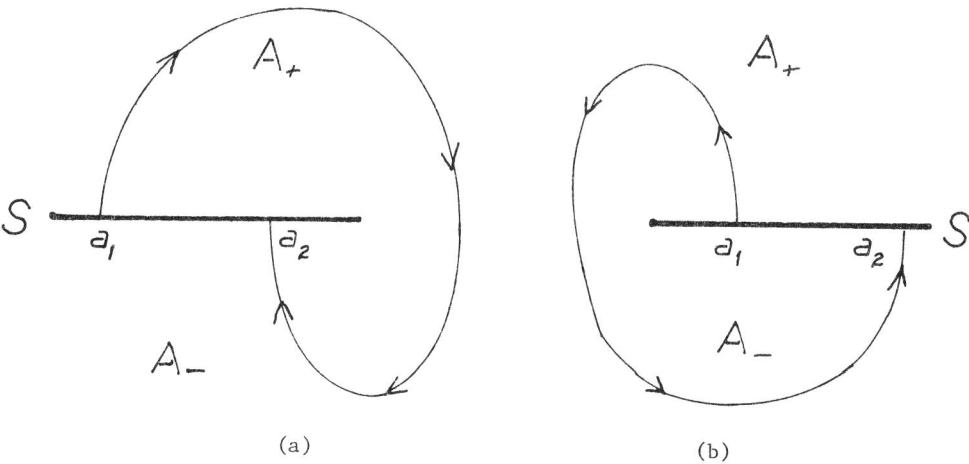

(a)   (b)

Figure (6.1)

(6.2) *Corollary:*

Suppose S is an interval transverse to the flow ϕ on $\mathbb{R}^2$, and $a_1 \neq a_2 \in S$ satisfy

(i) $a_2 \in 0_+(a_1)$
(ii) $0(a_1,a_2) \cap S = \emptyset$.

Pick the order on S so $a_1 < a_2$.
Then no ϕ-orbit crosses $S(a_1,a_2)$ twice, and one of the following holds:

(a) for every $z \in S$ with $z > a_1$, $0_+(z)$ is
     bounded, and if $0_-(z)$ is unbounded it meets
     $S[a_1,a_2]$.

or   (b) for every $z \in S$ with $z < a_2$, $0_-(z)$ is
     bounded, and if $0_+(z)$ is unbounded it meets
     $S[a_1,a_2]$.

Proof:

Any orbit crossing $S(a_1,a_2)$ goes from A_- to A_+ and can subsequently never return to $S(a_1,a_2)$. Note that for every $z \in S$, $z > a_1$ implies $0_+(z) \subset A_+$ while $z < a_2$ implies $0_-(z) \subset A_-$. If furthermore $z \notin S[a_1,a_2]$ then the initial orbit segments in both directions belong to

$A_{\pm}$. Since orbits can escape $A_{\pm}$ only via $S(a_1,a_2)$, we obtain (a) if A_+ is bounded and (b) if A_- is bounded. $\square$

(6.3) Corollary:

If $a_1 \neq a_2 \in S$ (transverse) with $a_2 \in \mathcal{O}_+(a_1)$, then $S[a_1 a_2] \cap \Omega(\phi) = \emptyset$.

Proof:

It suffices to prove (6.3) assuming $\mathcal{O}(a_1,a_2) \cap S = \emptyset$. Clearly (6.2) implies $S(a_1,a_2)$ wandering. To show $a_i \not\in \Omega(\phi)$, note that if $z_1 \in S$ is near a_1 , then the next intersection of $\mathcal{O}_+(z)$, with S is at z_2 near a_2 . Only one of the points z_i (i=1,2) can belong to $S(a_1,a_2)$. Now by (6.1) $\mathcal{O}_-(z_1) \subset A_-$ and $\mathcal{O}_+(z_2) \subset A_+$, so $S \cap \mathcal{O}_-(z)$ and $S \cap \mathcal{O}_+(z)$ are contained in distinct components of $S \setminus S[a_1,a_2]$. In particular, $\mathcal{O}_+(z_2)$ contains no points of S near a_1 , implying that a_1 (and hence a_2) wanders. $\square$

(6.4) Corollary:

If $\dot{\phi}(p) \neq 0$, there exists a transversal S intersecting the ϕ-orbit-closure of p only at p .

Proof:

Suppose a given section S intersects $\mathcal{O}(p)$ at a point q other than p . Applying (6.3) with $a_i = p$ and q respectively, we conclude that p wanders. Hence a wandering neighborhood of p intersects S in a section as required. $\square$

(6.5) Corollary:

If either limit set of $x \in \mathbb{R}^2$ under ϕ contains a non-restpoint, then x is periodic or wandering. In particular, each limit set of $x \in \Omega(\phi) \setminus Per(\phi)$ is either empty or consists of restpoints.

Proof:

Assume $x \not\in Per(\phi)$, and $p \in \omega(x)$, $\dot{\phi}(p) \neq 0$. By (6.4) there is a transversal S with $S \cap \text{clos } \mathcal{O}(p) = \{p\}$. If $x_1 \neq x_2$ are successive intersections of $\mathcal{O}_+(x)$ with S , then by (6.3) x_1 , hence x , wanders. The reformulation in the second part of (6.5) is immediate. $\square$

(6.5a) Remark:

If $x \in \Omega(\phi)$ and $\omega(x)$ or $\alpha(x)$ contains a hyperbolic restpoint or periodic orbit, then the local structure of hyperbolic orbits (in particular, the argument for (3.5 ii)) shows that this limit set contains nothing else. Furthermore, if $x \in \Omega(\phi) \setminus Per(\phi)$ and all fixedpoints in $\omega(x)$ or $\alpha(x)$ are hyperbolic, they must be fixed saddles, with x on one of their separatrices.

The next sequence of lemmas concerns escape to infinity.

(6.6) *Lemma*:

Suppose S *is a transverse section and some semiorbit of* $x \in S$ *escapes to infinity without crossing* S *again. Suppose some orbit crosses* S *twice (at* $y_+ \in \mathcal{O}_+(y_-)$, $y_+, y_- \in S$). *Pick an order on* S *so that* $y_- \leq x$. *Then:*

    (i) If $y_+ \leq y_-$, *then* $z \leq y_-$ *implies* $\mathcal{O}_+(z)$ *bounded. (Fig. 6.2 a)*

    (ii) If $y_- \leq y_+ \leq x$, *then* $z \leq y_+$ *implies* $\mathcal{O}_-(z)$ *bounded. (Fig. 6.2 b)*

    (iii) If $y_- \leq x \leq y_+$, *then*

        (a) $\mathcal{O}_-(x)$ *unbounded implies every* $z \geq y_-$ *has* $\mathcal{O}_+(z)$ *unbounded. (Fig. 6.2 c)*

        (b) $\mathcal{O}_+(x)$ *unbounded implies every* $z \leq y_+$ *has* $\mathcal{O}_-(z)$ *unbounded. (Fig. 6.2 d)*

Proof:

Note that the successive points of intersection of $\mathcal{O}(y_-)$ with S are monotone with respect to the order $\leq$. Thus, it suffices to prove (6.6) assuming $\mathcal{O}(y_-, y_+) \cap S = \emptyset$.

Apply (6.2) with $a_i = y_\pm$, reversing time direction if necessary, and noting which of the components $A_\pm$ contains the unbounded semiorbit of x. $\square$

(6.7) *Lemma*:

Suppose $a \neq b \in S$ *(transverse) are points which both escape to infinity in the same time direction without re-crossing* S. *(Say for definiteness that* $\alpha(a) = \emptyset = \alpha(b)$, $\mathcal{O}_-(a) \cap S = \emptyset = \mathcal{O}_-(b) \cap S$.) *Then for any pair* $x_\pm \in S$ *such that* $x_+ \in J^+(x_-)$, x_+ *does not belong to* $S(a,b)$. *If furthermore* $\mathcal{O}_+(x_+)$ *is unbounded, then* x_+ *separates* x_- *from both* a *and* b *in* S, *with the exception that* x_+ *may equal* a *or* b *(see fig. 6.3).*

Proof:

The complement of $L = \mathcal{O}_-(a) \cup S[a,b] \cup \mathcal{O}_-(b)$ has two components (which we again denote $A_\pm$) each invariant in one time direction. Note that for every $z \in S$, $\mathcal{O}_+(z) \subset A_+$, while for $z \in S(a,b)$, $\mathcal{O}_-(z) \subset A_-$. Since x_+ and x_- are prolongationally related, $\mathcal{O}_-(x_+)$ and $\mathcal{O}_+(x_-)$ must lie in the same component of $\mathbb{R}^2 \setminus L$, which must be A_+. This guarantees that $x_+ \notin S(a,b)$. Furthermore, if $\mathcal{O}_+(x_+)$ is unbounded, it cannot cross S twice, for otherwise (6.6) with $x = y_- = x_+$ would guarantee $\mathcal{O}_+(x_+)$ bounded. Now applying (6.6) to points $x_k \to x_-$ and $x_k' \in \mathcal{O}_+(x_k) \to x_+$ we see

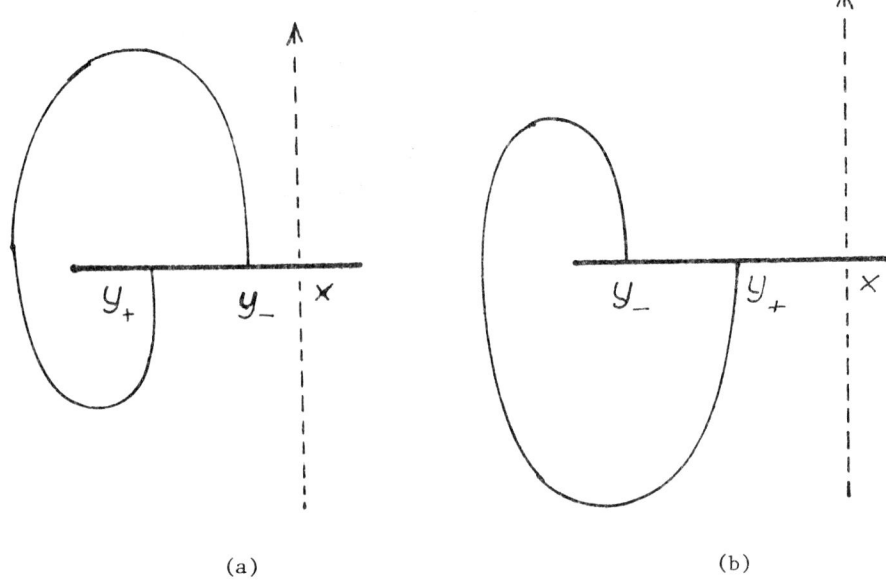

(a)   (b)

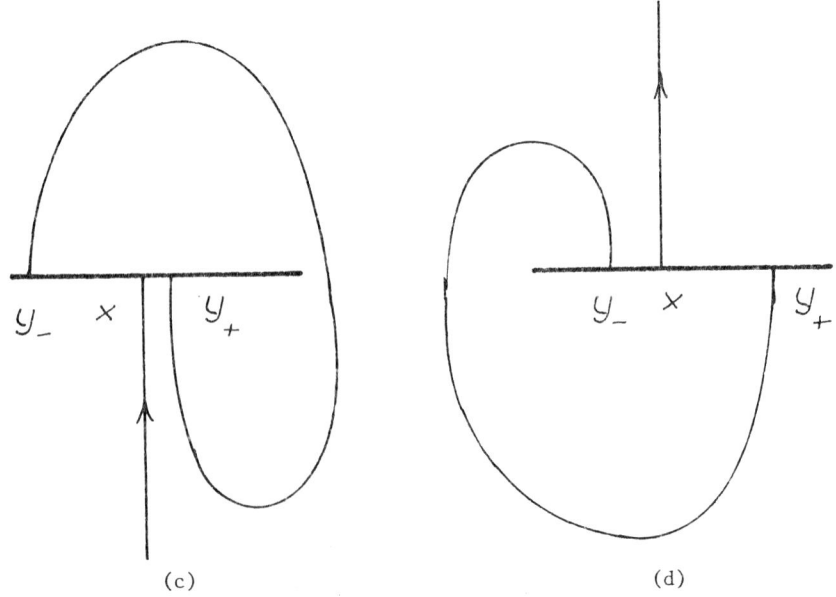

(c)   (d)

Figure 6.2

that x_- cannot lie between x_+ and either a or b. Thus, x_+ separates x_- from both a and b. ☐

By contrast with (6.7), where two orbits escape to infinity in the same direction, the following lemma treats escape to infinity in opposite directions:

(6.8) *Lemma:*

Suppose $a \neq b \in S$ *(transverse) escape to infinity in opposite directions without re-crossing* S, *say* $\alpha(a) = \emptyset = \omega(b)$ *and* $O_-(a) \cap S = \emptyset = O_+(b) \cap S$. *Pick an ordering with* $a < b$. *Suppose some pair* $x_\pm \in S$ *satisfies* $x_+ \in J^+(x_-)$. *If either* $O_+(x_+)$ *or* $O_-(x_-)$ *is unbounded, then* $x_- \leq b$ *implies* $x_+ \leq a$, *while* $x_- \geq b$ *implies* $x_+ \geq a$ *(see fig. 6.4).*

Proof:

Again, let $A_\pm$ be the components of the complement of $L = O_-(a) \cup S[a,b] \cup O_+(b)$. For $z < b$, $O_+(z) \subset A_+$, while for $z > a$, $O_-(z) \subset A_-$. If $x_+ < a$ and $x_- > b$, then there exist orbits starting near x_- which cross S twice, first in $S(a,b)$ and then near x_+. By (6.6) this implies $O_-(z)$ bounded for every $z \geq b$ and $O_+(z)$ bounded for $z \leq a$. In particular, neither $O_+(x_+)$ nor $O_-(x_-)$ can be unbounded. ☐

Finally, we deal with a single orbit escaping to infinity in both time directions:

(6.9) *Lemma:*

If $x \in S$ *(transverse) escapes to infinity in both time directions and* $y \in S \cap J^+(x)$, *then every point* $z \in S$ *on the same side of* x *as* y *has* $O_+(z)$ *bounded.*

Proof:

Note that by (6.2), $O(x) \cap S = \{x\}$. Suppose $x < y$. Then (6.6) applied to $p_k \to x$ and $p'_k \in O_+(p_k) \to y$ gives (6.9). ☐

The main application of these lemmas will be in the proof of theorem C. In this proof, one of our main concerns is the generic absence of any *triple*, by which we mean points p_i, $i=1,2,3$, such that $O_+(p_j)$ and $O_-(p_{j+1})$ form a saddle at infinity for $j = 1,2$. Note that the middle point p_2 escapes to infinity in both time directions. Thus, $O(p_2)$ separates the plane. We will call the triple a *one-sided* (resp. *two-sided*) *triple* if p_1 and p_3 lie on the same (resp. opposite) side of $O(p_2)$. The study of triples (and their generic non-existence) in the completely unstable planar case was carried out in [Kr 2].

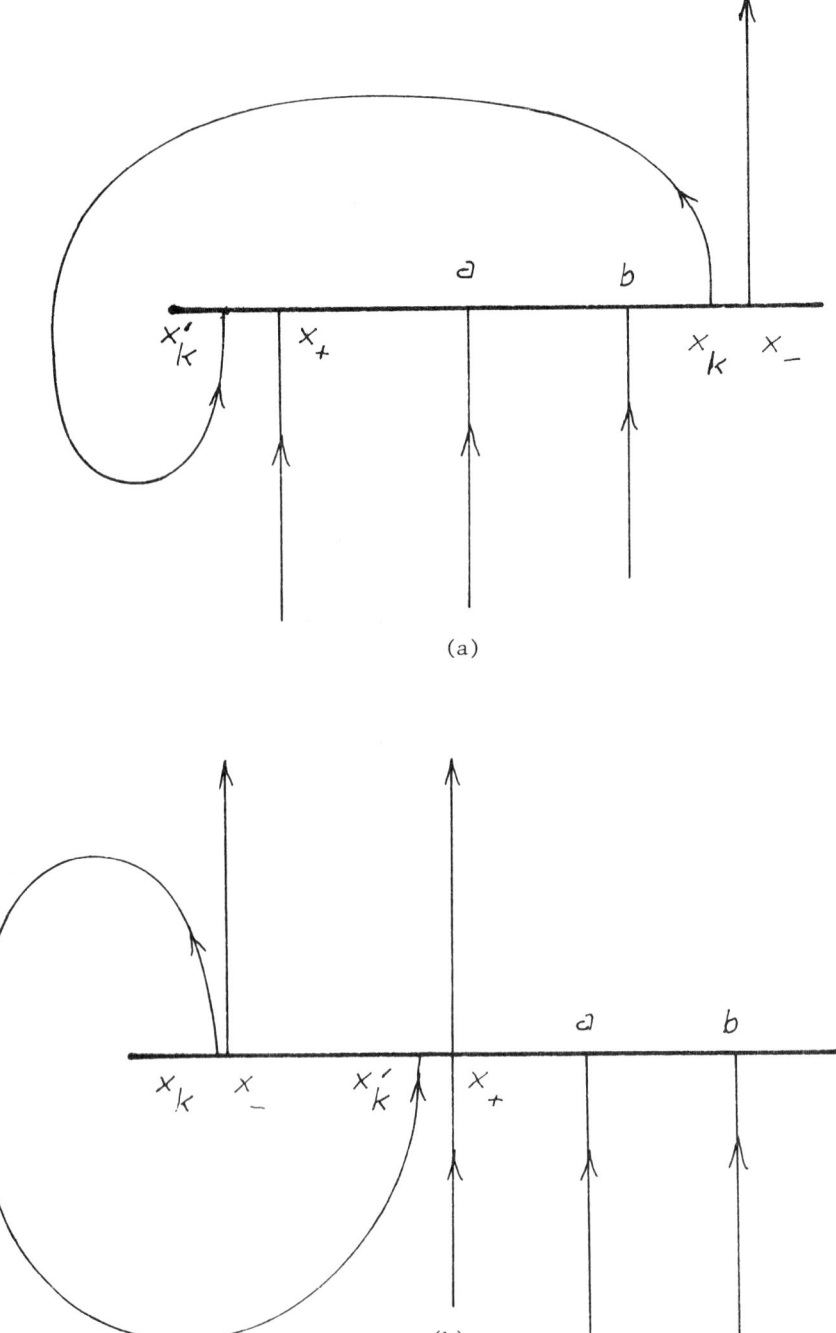

(a)

(b)

Figure 6.3

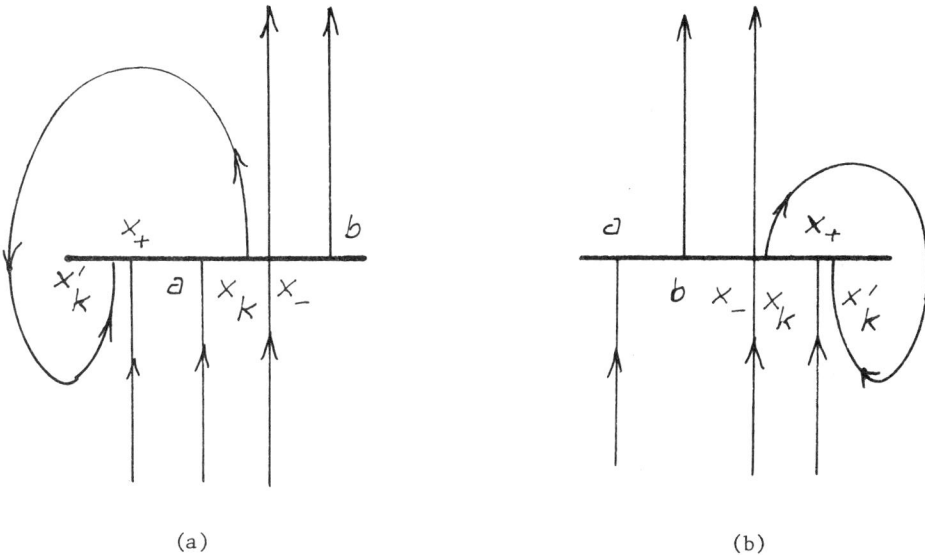

(a) (b)

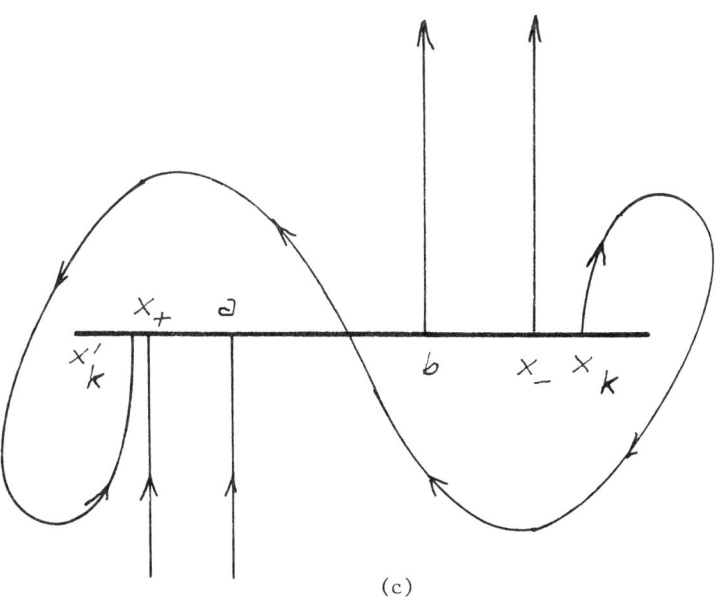

(c)

Figure 6.4

(6.10) *Lemma:*

   If all three points p_i, $i=1,2,3$ of a triple lie on a single transverse interval S, then the triple is two-sided.

Proof:

   Apply (6.9) to $y = p_2$, $x = p_3$ to conclude that p_1 lies on the
opposite side of p_2 from p_3. $\square$

   When $O_+(p_+)$ and $O_-(p_-)$ form a saddle at infinity, we know there
exist $x_k \to p_-$ with $x_k' \in O_+(x_k) \to p_-$; the technical lemma below allows us
to pick x_k and x_k' so that x_k' is the first point on $O_+(x_k)$ near p_-.

(6.11) *Lemma:*

   Suppose $p_- \in J^+(p_+) \setminus clos\ O_+(p_+)$, where $p_\pm$ are not restpoints, and
suppose S_+ is a compact transverse interval through p_+, disjoint from
$clos\ O_-(p_-)$. Then there exists a compact transverse interval S_- through
p_- and points $x_k \to p_+$ in S_+ such that the first intersections x_k' of
$O_+(x_k)$ with S_- satisfy $x_k' \to p_-$.

Proof:

   Pick any compact section S at p_- disjoint from $clos\ O_+(p_+)$ and
points $x_k \to p_+$, $y_k \in O_+(x_k) \cap S$ with $y_k \to p_-$. Let x_k' be the first
point of $O_+(x_k)$ on S. Note that (6.2) with $a_1 = x_k'$ and $a_2 \in y_k$
precludes the possibility that y_k lies between x_k' and p_-. If x_k' lies
between y_k and p_- for all large k, (or for a subsequence) then we are
done with $S = S_-$.

   Suppose for some k that x_k' and y_k lie on opposite sides of p_- in
S (see fig. 6.5). Then by (6.2) no orbit crosses $S(x_k', y_k)$ twice. If we
have S_- a slightly smaller closed neighborhood of p_- in S, then any
orbit crosses S_- at most once, but $y_k \to p_-$ guarantees that $y_k \in S_-$ for
k large. Thus $y_k = k_k'$ is the only (hence first) point of $O_+(x_k)$ in S_+,
and we are done. $\square$

   We close this section with several technical lemmas concerning the possibility of achieving certain phenomena under c^r perturbation. We gather
these together here so as not to clutter the global picture when we prove
our genericity results in section 7. The first lemma is a reformulation of
the local perturbation argument used by Peixoto in [Pe 2].

(6.12) *Lemma:*

   Given a flowbox F for ϕ, a point $p \in int\ S_+$, and a c^r-neighbor-
hood U of ϕ, there exists a neighborhood $\widetilde{S}_+$ of p in S_+ and a flow-
box $\widetilde{F} \subset F$ with entrance set $\widetilde{S}_+$ (and corresponding exit set $\widetilde{S}_- \subset S_-$)
such that for any pair of points $q_\pm \in \widetilde{S}_\pm$ there exists a flow ϕ satisfying

(i) $\psi \ \varepsilon \ U$

(ii) $\dot{\psi} = \dot{\phi} \ \ off \ \ F$

(iii) $q_- \ \varepsilon \ 0(q_+,\psi)$, and $0_\psi[q_+,q_-] \subset F$.

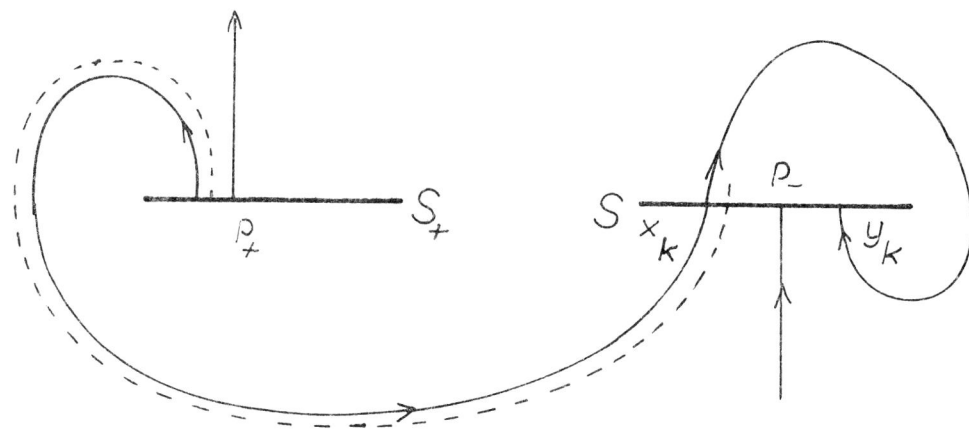

Figure 6.5

Proof:

Pick a C^∞ function $f : \mathbb{R}^2 \to [0,1]$ which vanishes off F and is positive at every point of int F. For $u \ \varepsilon \ \mathbb{R}^2$, let

$$Y_u(x) = \dot{\phi}(x) + uf(x) \ V$$

where V is a vectorfield perpendicular to $\dot{\phi}$ on F. There exists $\varepsilon > 0$ such that the flow ψ_u generated by Y_u belongs to U for all $u \ \varepsilon \ [-\varepsilon, \varepsilon]$, and ψ_u satisfies (ii). Furthermore, for $u \neq 0$, the vectorfield Y_u is transverse to $\dot{\phi}$ on int F.

For $q \ \varepsilon$ int S_+ and $|u|$ small, we can define $g(u,q)$ as the first point of $0_+(q,\psi_u)$ on S_- (ie, $g(u, \cdot)$ is the Poincaré map of ψ_u in the sense of (4.2)). It is easy to see that g is a continuous function of two variables and for q fixed $g(u,q)$ is strictly monotone u. Thus, for $\delta > 0$ small and q near p, the set

$$G(\delta,q) = \{g(u,q) \ | \ |u| < \delta\}$$

varies continuously with q, so that given $0 < \delta < \varepsilon$ we can find $q_1 < g(0,p) < q_2$ and a neighborhood V of p in S_+ such that for every $q \ \varepsilon \ V$, $G(\delta,q)$ includes $[q_1,q_2]$. If we pick $p_i = S_+ \cap 0_-(q_i,\phi)$, i=1,2 then the flowbox $\widetilde{F}$ defined by

$$\widetilde{S}_+ = S_+[p_1,p_2], \quad \widetilde{S}_- = S_-[q_1,q_2]$$

satisfies (iii) for appropriate $\psi_u \ \varepsilon \ U$. $\square$

The following is a kind of "closing lemma" for prolongations:

(6.13) _Lemma_:

   Suppose $x_2 \in J^+(x_1, \phi) \setminus \text{clos } 0_+(x_1, \phi)$, _where_ $\dot{\phi}(x_i) \neq 0$ (i=1,2).
Given disjoint neighborhoods V_i _of_ x_i _and a_ C^r _neighborhood_
U _of_ $\phi (r \geq 0)$, _there exists_ $\psi \in U$ _such that_

   (i) $\dot{\psi} = \dot{\phi}$ _off_ $V_1 \cup V_2$

   (ii) $x_2 \in 0_+(x_1, \psi)$.

Proof:

   Pick $S_i \subset V_i$ compact transverse sections at x_i as in (6.11)
$(p_+ = x_1, \ p_- = x_2)$. Pick $F^i \subset V^i$ flowboxes with S_1 (resp. S_2) the
entrance (resp. exit) set of F_1 (resp. F_2), and apply (6.12) to F_i to ob-
tain $\widetilde{F}_i$, with $x_i \in \widetilde{S}_i \subset S_i$. Let S_i^* denote the opposite transverse edge
of F_i. By (6.11), there are points $p_i \in S_i^*$ such that p_2 is the point
of $0_+(p_1, \phi)$ on S_2, and by (6.12) there exist flows $\dot{\psi}_i \in U$, i=1,2 such
that $\dot{\psi}_i = \dot{\phi}$ off F_i and $p_1 \in 0_+(x_1, \psi)$, $p_2 \in 0_-(x_2, \psi_2)$ (and
$0_{\psi_i}[x_i, p_i] \subset F_i)$. But then the equations

$$\dot{\psi}(x) = \dot{\psi}_i(x) \quad \text{for} \quad x \in V_i$$

$$\dot{\psi}(x) = \dot{\phi}(x) \quad \text{for} \quad x \notin V_1 \cup V_2$$

define a flow $\psi \in U$ with $x_2 \in 0_+(x_1, \psi)$. □

(6.14) _Corollary_:

   Suppose $0(x_2, \phi)$ _escapes to infinity in both directions and_
$0_+(x_1, \phi), 0_-(x_2, \phi)$ _form a saddle at infinity. Given neighborhoods_ V_i _of_
x_i (i=1,2) _and a_ C^r-_neighborhood_ U _of_ ϕ, _there exist_ $\psi \in U$ _such that_

   (i) $\dot{\psi} = \dot{\phi}$ _off_ V_2

   (ii) $0_+(x_i, \psi) = 0_+(x_i, \phi)$ i=1,2

   (iii) $0_-(x_2, \psi)$ _intersects_ V_1

   (iv) _if_ $x_2 \notin J^+(x_2, \phi)$ _and_ x_3 _is a point with_ $x_3 \in J^+(x_2, \phi)$,
         then also $x_3 \in J^+(x_2, \psi)$.

Proof:

   By (6.4) there exists a transverse section $S_2 \subset V_2$ through x_2 intersecting the orbit of x_2 only at x_2. Form a flowbox $F \subset V_2$ with exit
set S_2. There exist $p_i \to x_1$ and $p_i' \in 0_+(p_i, \phi) \cap S_+$ (the first such
intersection, by (6.11)) so that $p_i' \to 0_-(x_2, \phi)$. By (6.12) for i large
there exists $\psi \in U$ with $\dot{\psi}_i = \dot{\phi}$ off F and $x_2 \in 0_+(p_i', \psi_i)$, so we have
(i)-(iii). If $x_2 \notin J^+(x_2, \phi)$, we can pick S_2 so that no ϕ-orbit crosses

S_2 twice, and hence any $x_3 \in J^+(x_2, \phi)$, still belongs to $J^+(x_2, \psi_i)$. $\square$

(6.15) *Corollary*:

Suppose as in (6.14) that $0(x_2, \phi)$ escapes to infinity in both directions and $0_+(x_1, \phi)$, $0_-(x_2, \phi)$ form a saddle at infinity. Suppose further
that x_1 and x_2 belong to a single transverse section S. Then there
exist a nonempty C^r-open set V containing ϕ in its C^r-closure $(r \geq 1)$
and points

$$x_1(\psi) \leq x_2(\psi) < x_3'(\psi) < x_4(\psi) \in S$$

(with $x_1(\psi) < x_2(\psi)$ if $x_1 \neq x_2$), varying continuously with $\psi \in V$ such
that

   (i) $0_-(x_4(\psi), \psi)$ escapes to infinity

   (ii) $0_+(x_i(\psi), \psi)$ escapes to infinity for $i=1,2$

   (iii) for every $y \leq x_3'( )$ in S, $0_-(y)$ is bounded.

Proof:

   (See fig. 6.6a.) Apply (6.14) so that $0_-(x_2, \psi_1)$ passes near x_1 at
$x_2'' \in S$. Some point x_3 between x_2'' and x_1 crosses S again at x_3', so
that x_2 lies between x_1 and x_3'. (Fig. 6.6b) These choices can be made
continuously on some neighborhood U_1 of ψ_1. Since ψ_1 is arbitrarily
C^r-near ϕ, the union of the corresponding U_1 forms an open set V with
ϕ in its closure for which (i) and (ii) hold. Then (iii) follows from
(6.6). $\square$

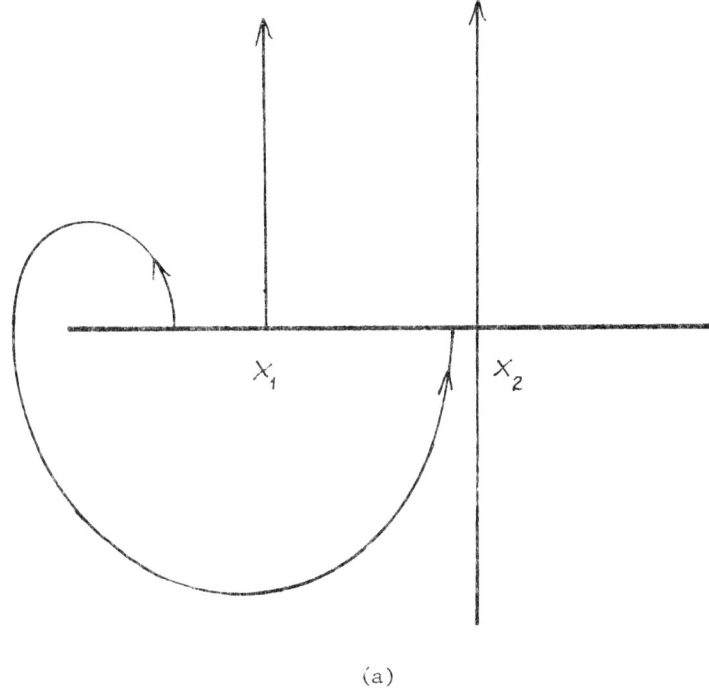

(a)

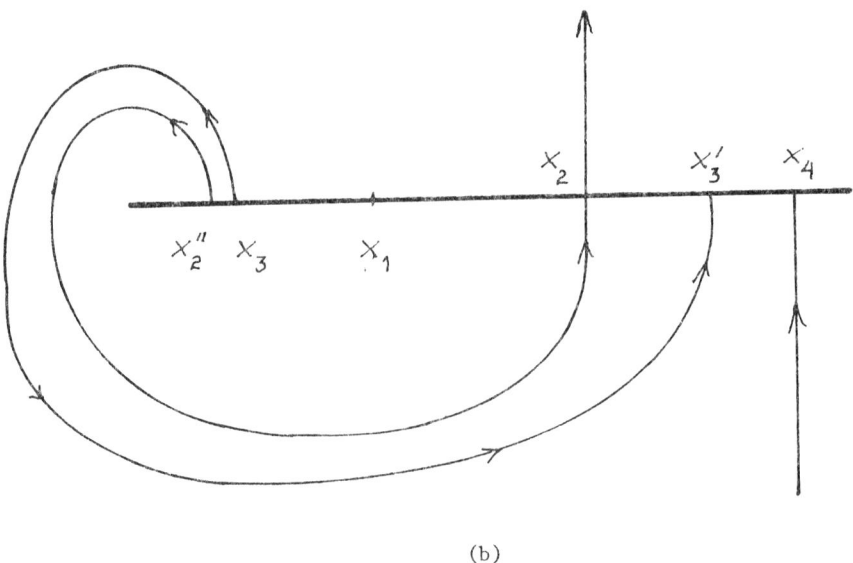

(b)

Figure 6.6

7. *Generic properties of flows in the plane*

In this section, we consider the genericity results (1.4-6) stated in §1. The first of these, the Kupta-Smale theorem (1.4) is well-known, and has been proved for any open manifold by Peixoto [Pe 3]. We will not prove (1.4) here; however, we will examine some aspects of this result to glean further information which will be useful to us. The second result, generic absence of oscillating orbits (1.5), was proved in [Ko], but we need to reproduce the heart of this proof for other purposes; (1.5) is a part of (7.6). The bulk of our energy will be devoted to theorem C (1.6), the genericity of the condition $W^-(\phi) \cap W^+(\phi) \subset Per(\phi)$.

Recall that a property is (strongly) C^r-*generic* if there exists a residual set (countable intersection of dense open sets) of flows in the (strong) C^r topology every element of which has the specified property. It will be useful to us to know that certain generic properties actually hold on a dense open set.

We begin with hyperbolicity. Standard arguments (see [Pe]) in the compact case show that, given a compact region G, the vectorfields for which every fixedpoint in G is interior to G and hyperbolic form a C^1 compact-open set, concentrated on G (or a neighborhood of G). Covering any open surface M with a locally finite family of compact regions, and using C^r-density ($r \geq 1$) we can show

(7.1) <u>Lemma</u> ([Pe 3]):

For any surface M, *the set* $\bar{F}(M)$ *of vectorfields on* M *for which every restpoint is hyperbolic is a (strong)* C^r-*dense open set for* $r \geq 1$.

By contrast, the hyperbolicity of periodic orbits is not a C^r-open condition. Even on a compact region, a flow with a cycle of saddle-connections can be perturbed to yield non-hyperbolic periodic orbits with long periods. Similarly, the examples in fig. 2.7 can be perturbed on a compact flowbox so that some of the large-amplitude periodic orbits become non-hyperbolic.

In a similar way, in the presence of non-hyperbolic periodic orbits, the non-existence of connections between fixed saddles (even on a compact region) is not an open condition. For example, a flow with a periodic orbit that attracts an unstable separatrix (of some fixed saddle) from one side and repels a stable separatrix (of another fixed saddle) from the other side (fig. 7.1) can be perturbed to create a new saddle connection.

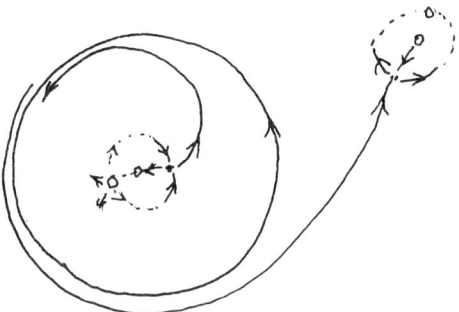

Figure 7.1

However, the combination of these two conditions, at least on a compact region, does define a C^r-open set $(r \geq 1)$. To prove this, we combine (4.3-4.4) with the following technical lemma, which is specific to the plane:

(7.2) Lemma:

   Given a compact region $C \subset \mathbb{R}^2$, suppose the flow ϕ satisfies:

    (i)   every restpoint in C is interior to C and hyperbolic

    (ii) every periodic orbit entirely contained in C is interior to C and hyperbolic

   (iii) there is no saddle connection entirely contained in C .

Then:

    (iv) there are finitely many restpoints and periodic orbits entirely contained in C

    (v)   if a stable (resp. unstable) separatrix $W^{\pm}(p)$ (p a fixed saddle) is entirely contained in C , then its α-(resp. ω-) limit set is a single fixed or periodic source (resp. sink).

Proof:

   Note that a sink or source has a neighborhood intersecting no other orbits in $\mathrm{Per}(\phi)$. Hence an accumulation point of periodic orbits in C must be either a fixed saddle or a point in $\Omega(\phi) \setminus \mathrm{Per}(\phi)$ whose orbit is entirely contained in C. By an argument as in (3.5), if a sequence of periodic orbits accumulate on a fixed saddle, they also accumulate on one of its separatrices. Thus, in either case, we obtain an orbit in $\Omega(\phi) \setminus \mathrm{Per}(\phi)$ entirely contained in C. But remark (6.5a) shows that any such orbit must be a saddle connection in C .   This contradiction proves (iv).

In a similar way, the α- and ω-limit of any separatrix $\overline{W}^+(p)$ entirely contained in C consists of non-wandering orbits entirely contained in C , hence every such limit set consists of a single orbit in Per(φ) . But by assumption, the α- (resp. ω-) limit of a stable (resp. unstable) separatrix in C cannot be a saddle, so we have (v). ☐

Lemmas (4.3-4) and (7.2) allow us to extend (7.1), at least on a compact region in the plane:

(7.3) *Proposition:*

   Given a compact region $C \subset \mathbb{R}^2$, denote by $\mathscr{G}(C)$ the flows φ for which

      (i) every orbit of Per(φ) entirely contained in C is
            hyperbolic and interior to C

     (ii) there are no saddle connections entirely contained
          in C .

   Then $\mathscr{G}(C)$ is open in the C^r compact-open (and hence strong) topology .

Proof:

Fix $φ \in \mathscr{G}(C)$.   By (4.3) and (4.4), together with (7.2iv), there are neighborhoods $V \subset \text{int } C$ of the fixed and periodic φ-orbits contained in C and a C^r compact-open neighborhood U of φ (concentrated on C) such that for $\psi \in U$, V contains no new periodic or fixed orbits, all the "old" ones remain hyperbolic, and $\dot{\psi} \neq 0$ on $C \setminus V$. Suppose now that $φ_n \to φ$ (C^r on C) . For n large, $φ_n \in U$, and so it has no fixed points outside V , and by (4.4) it has no saddle connections. Thus, if $φ_n \notin \mathscr{G}(C)$, this must result from the existence of $φ_n$-periodic orbits contained in C but not contained in any V . The set of accumulation points as $n \to \infty$ for these new $φ_n$-periodic orbits form a closed φ-invariant subset of C , and thus contains a nonempty minimal set. By (6.4) or (6.5), such a set must be a single orbit of Per(φ) , contained in C . Since a periodic orbit of $φ_n$ cannot intersect a neighborhood of a sink or source of φ (since this is similarly a neighborhood of a sink or source of $φ_n$), it follows that this minimal set is a saddle. But it is easy to see that in this case some unstable separatrix of this saddle is also an accumulation point of $φ_n$-periodic orbits as $n \to \infty$, as is its ω-limit, which must be a sink. Thus we are forced to have $φ_n$-periodic orbits intersecting but not contained in a neighborhood of some sink for φ ; this contradiction gives $φ_n \in \mathscr{G}(C)$ for n large . ☐

We note in passing that (7.3) could be used (via nested compact regions) to prove (1.4) in the plane by a method differing slightly from that of Peixoto (who considers an increasing upper bound on the periods of circular orbits).

On the other hand, we do not know if the set of flows for which all orbits in
Per(ϕ) are hyperbolic (which is dense but not itself open) contains a C^r
dense-open set, for M an open surface (even $M = \mathbb{R}^2$). However, this is of
limited interest for the purposes of this paper.

We turn now to the "local" result below involving separatrices of fixed
saddles. This is also the local part of the proof of (1.5) in [Ko].

(7.4) _Lemma_:

Given a compact region $C \subset \mathbb{R}^2$, there exists a set $\mathcal{G}0(C)$ which is
strongly C^r dense and open such that if $\phi \in \mathcal{G}0(C)$ and $\alpha(x, \phi) \subset C$ then
$0_+(x, \phi)$ does not oscillate.

Proof:

Using (7.3), we can restrict attention to the dense-open set $\mathcal{G}(C)$.
Thus by (6.5) any α-limit set contained in C is either a single source in
Per(ϕ) interior to C or a fixed saddle interior to C. By (4.3-4), there
are finitely many such orbits and they all vary continuously in int C.
Thus, by restricting attention to a neighborhood of a given flow ϕ we can
consider one source or saddle at a time.

We begin by showing $\mathcal{G}0(C)$ dense. Suppose then that $\phi \in \mathcal{G}(C)$ has
$0_+(x, \psi)$ oscillating and $\alpha(x, \phi) \subset C$ and pick $V \subset \mathcal{G}(C)$ a C^r-neighbor-
hood of ψ. Since $0_+(x, \psi)$ oscillates, $\omega(x, \phi)$ contains a non-fixed-
point $y \notin C$; take S_- a transverse section through y, and build a flowbox F with exit set S_- such that $F \cap C = \emptyset$. Use (6.12) to find a flow-
box $\widetilde{F} \subset F$ with $y \in S_-$ such that every pair $z_\pm \in \widetilde{S}_\pm$ can be joined by a
ψ-orbit in F for some $\psi \in V$.

Since $y \in \omega(x, \phi)$, there exist $x_1, x_2 \in \widetilde{S}_+ \cap 0_+(x, \phi)$ such that x_2
is the first point of $0_+(x_1, \phi)$ on S_+, and x_2 lies between x_1 and y.
Then
$$L = 0_\phi[x_1, x_2] \cup S[x_1, x_2]$$
bounds a negatively invariant region A_- containing $0_-(x, \phi)$. Let $x_1^- \in \widetilde{S}_-$
be the point at which $0_+(x_1, \phi)$ leaves $\widetilde{F}$. Using (6.12) find $\psi \in V$ such
that

(i) $\dot{\psi} = \dot{\phi}$ off F

(ii) $x_1^- \in 0_+(x_2, \psi)$, and $0[x_2, x_1^-] \subset F$.

Now, $0_+(x_2, \psi)$ intersects $\widetilde{S}_+$ at x_1, and
$$L' = 0_\psi[x_2, x_1] \cup \widetilde{S}[x_2, x_1]$$
bounds a positively invariant region that includes $A_- \setminus F$, and in particular
every ψ-orbit with α-limit near $\alpha(x, \phi)$ is positively bounded, and hence

does not oscillate.

This property holds for ψ (so $\psi \varepsilon \mathcal{G}0$(C)) but also for all ψ near ψ, so that $\mathcal{G}0$(C) can be chosen to be dense and open. $\square$

As we noted in §2, the condition of theorem C, $\overline{W}^-(\phi) \cap W^+(\phi) = \emptyset$, has three parts:

1. No saddle connections

2. No fixed saddles for which some separatrix forms a saddle at infinity with another semi-orbit

3. No triples (in the sense of §6).

The Kupka-Smale theorem (1.4) shows that condition (1) is generic. To prove genericity of (2), we use (7.4) to establish the following "local" version:

(7.5) *Lemma:*

Given $C \subset \mathbb{R}^2$ a compact region, there is a C^r-dense open set $\mathcal{G}_1(C)$ $\subset \mathcal{G}0$(C) $(r \geq 1)$ of flows ϕ such that if $p \varepsilon C$ is a fixed saddle and $q \varepsilon C$ then neither branch of $W^{\pm}(p, \phi)$ can form a saddle at infinity with $O_{\pm}(q, \phi)$.

Proof:

We argue by contradiction, using $\overline{W}^-(p, \phi)$ and $O_-(q, \phi)$. If (7.5) is false, some open set $\mathcal{U} \subset \mathcal{G}0$(C) contains a dense subset V such that each $\psi \varepsilon V$ has a fixed saddle $p(\phi) \varepsilon C$ and a branch γ of $\overline{W}^-(p(\psi), \psi)$ which forms a saddle at infinity with $O_-(q(\psi), \psi)$ with some $q(\psi) \varepsilon C$. Using (4.3-4) and (7.3) we can assume $p(\phi)$ is defined to vary continuously with $\phi \varepsilon \mathcal{U}$, and that we can pick out one branch $\gamma(\phi)$ of $\overline{W}^-(p(\phi), \phi)$, whose initial compact segments vary continuously with $\phi \varepsilon \mathcal{U}$, such that when $\psi \varepsilon V$ then $\gamma(\psi)$ forms a saddle at infinity with $q(\psi) \varepsilon C$. Note that we assume no continuous dependence of $q(\psi)$ on ψ. Assuming these choices of $p(\phi)$, $\gamma(\phi)$ for $\phi \varepsilon \mathcal{U}$ and $q(\psi)$ for $\psi \varepsilon V$, we will produce a flow $\xi \varepsilon \mathcal{U}$ for which $\gamma(\xi)$ oscillates, contradicting $\mathcal{U} \subset \mathcal{G}0$(C).

To this end, pick $\psi_1 \varepsilon V$, so that $\gamma(\psi_1)$ and $O_-(q(\psi_1))$ form a saddle at infinity. Set $q_1 = q(\psi_1) \varepsilon C$, and pick $r_1 (\psi_1) \varepsilon \gamma(\psi_1)$ with $\| r_1 (\psi_1) \| > 1$. By (4.4), we can pick $r_1 (\phi) \varepsilon \gamma(\phi)$ varying continuously with ϕ near ψ_1 such that $\| r_1 (\phi) \| > 1$. Using (6.13), there exists ϕ_1 near ψ_1 such that $q_1 \varepsilon O_+(r_1 (\phi), \phi_1)$. Again using (4.4), we can pick $q_1(\psi)$ varying continuously with ψ near ϕ_1 such that $q_1(\psi) \varepsilon O_+ (r_1 (\psi), \psi)$.

Now, pick ψ_2 near ϕ_1 such that $\gamma(\psi_2)$ forms a saddle at infinity with some $q(\psi_2) \varepsilon C$. Set $q_2 = q(\psi_2)$, and pick $r_2 \varepsilon O_+(q_1(\psi_2), \psi_2) \subset \gamma(\psi_2)$ such that $\| r_2 \| > 2$. Again by (4.4) we can choose

$r_2(\phi) \in \mathcal{O}_+(q_1(\phi), \phi) \subset \gamma(\phi)$ varying continuously with ϕ near ψ_2 so that $\| r_2(\phi) \| > 2$.

Proceeding inductively, we find flows ϕ_n such that for ψ near ϕ_n there are points $q_i(\psi)$ and $p_i(\psi)$ varying continuously with ψ (near ϕ_n) such that

(i) $q_i(\psi) \in \mathcal{O}_+(r_i(\psi), \psi) \subset \gamma(\psi)$ for i=1, ..., n

(ii) $r_i(\psi) \in \mathcal{O}_+(q_{i-1}(\psi), \psi)$ for i=2, ..., n

(iii) $\| r_i(\psi) \| > i$ i=1, ..., n .

Then we find $\psi_{n+1} \in V$ near ϕ_n so that (i-iii) hold and $\gamma(\psi_{n+1})$ forms a saddle at infinity with some $q_{n+1} \in C$. Pick $r_{n+1}(\phi)$ varying continuously with ϕ near ψ_{n+1} and satisfying (ii) and (iii) for i=n+1; using (6.13) find ϕ_{n+1} with $q_{n+1} \in \mathcal{O}_+(r_{n+1}(\phi_{n+1}), \phi_{n+1})$, and pick $q_{n+1}(\psi)$ varying continuously with ψ near ϕ_{n+1} such that (i) holds. This is the inductive step.

It is clear that, since "near" can be defined arbitrarily at each stage, the sequence ϕ_n can be made to converge C^r-uniformly on compacta (but probably not in the strong C^r topology) to some $\xi \in U$. For this flow, (i-iii) hold for all n, so we have alternating points $q_i(\xi)$, $r_i(\xi) \in \gamma(\xi)$ with $q_i(\xi) \in C$ but $\| r_i(\xi) \| > i$. This shows $\gamma(\xi)$ oscillates, contrary to $\xi \in U \subset \mathcal{G}0(C)$ and so proves (7.5) . ☐

Combining the results above, we obtain

(7.6) _Proposition:_

There exists a C^r residual set $\mathcal{G}_1$ of flows ϕ on $\mathbb{R}^2$ satisfying:

(i)   every orbit in $Per(\phi)$ is hyperbolic

(ii) there are no saddle connections between finite saddles

(iii) if $\mathcal{O}_+(x, \phi)$ and $\mathcal{O}_-(y, \phi)$ form a saddle at infinity, then neither $\alpha(x, \phi)$ nor $\omega(y, \phi)$ contains any fixed saddles

(iv) there are no oscillating orbits.

For such flows,

(v)   any orbit in $\Omega(\phi) \setminus Per(\phi)$ escapes to infinity in both time directions

(vi) the ω- and α-limit of any point is either empty or consists of a single orbit in $Per(\phi)$.

Proof:

It is clear that $\mathcal{G}_1 = \bigcap_i [\mathcal{G}_1(c_i) \cap \mathcal{G}(c_i)]$, where $c_i \subset c_{i+1} \cdots$

are compact regions with $\underset{i}{\cup} \; C_i \; = \; \mathbb{R}^2$, satisfies (i)-(iii)

(iv) holds because by (6.1a) $\alpha(x, \phi)$ (resp. $\omega(x, \phi)$) is bounded, hence is contained in some C_i , and we apply (7.4) .

To prove (v), note that if $x \; \epsilon \; \Omega(\phi) \setminus Per(\phi)$ and x fails to escape to infinity in both time directions, then at least one limit set is nonempty and hence by (6.5) and (6.5a) x is a separatrix for some fixed saddle. Since there are no saddle connections, x escapes to infinity in the other direction. Suppose $x \; \epsilon \; W^-(p, \phi)$ for some fixed saddle and $x \; \epsilon \; \Omega(\phi)$. Then $x \; \epsilon \; J^-(x, \phi)$, and so by an argument as in (3.5) some $y \; \epsilon \; W^+(p, \phi)$ belongs to $J^+(x, \phi)$. Thus y also escapes to infinity, and $O_+(x, \phi)$, $O_-(y, \phi)$ form a saddle at infinity, contrary to (iii) .

To prove (vi), note that if $\omega(x, \phi) \neq \emptyset$ and does not consist of a single periodic orbit, they by (v) $O_+(x, \phi)$ oscillates. $\square$

Even, though we showed in (3.2a) that the condition $\Omega(\phi) = Per(\phi)$ is a consequence of hypotheses which we now want to show generic, it will be convenient to establish the genericity of this condition directly at this stage, to simplify our further arguments. This is done in the following:

(7.7) _Lemma_:

$\quad$ _Given a compact arc_ $S \subset \mathbb{R}^2$, _set_ U _the_ C^r -_open set_ $(r \geq 1)$ _of flows for which_ S _is a transverse section. For a dense open subset_ $V \subset U$,
$\psi \; \epsilon \; V$ _implies_

$$S \cap \Omega(\phi) \subset Per(\phi) .$$

Proof:

$\quad$ Pick $\phi_0 \; \epsilon \; \mathcal{G}_1 \cap U$. If $p \; \epsilon \; S \cap \Omega(\phi_0)$ is not periodic, then by (7.6v) it escapes to infinity in both directions, so that $O_+(p, \phi)$, $O_-(p, \phi)$ form a saddle at infinity. Using (6.15) (with $x_1 = p = x_2$), we obtain an open set $U_0 \subset U$ of flows near ϕ_0 for which there exist $x_1(\phi) < p < x_3'(\phi)$ such that $O_+(x_1, \phi)$ escapes to infinity and every $z < x_3'(\phi)$ has $O_-(z, \phi)$ bounded. In particular, if $\psi \; \epsilon \; \mathcal{G}_1 \cap U_1$, then for every $z < x_3'(\psi)$ either $z \; \epsilon \; Per(\psi)$ or z wanders.

$\quad$ Suppose now that some $\psi \; \epsilon \; \mathcal{G}_1 \cap U_1$ has $p' \; \epsilon \; S \cap [\Omega(\psi) \setminus Per(\psi)]$. Then $p' > x_3'(\psi)$ and a similar argument gives $U_2 \subset U_1$ on which we have $y_3'(\phi) < p' < y_1(\phi)$, $O_+(y_1, \phi)$ escaping to infinity and every semi orbit $O_-(z, \phi)$ for $z > y_3'(\phi)$ bounded.

$\quad$ Finally, for $\phi \; \epsilon \; U_2 \cap \mathcal{G}_1$, if $q \; \epsilon \; S \cap \Omega(\phi)$, then

$\quad$ (i) if $q \notin S[x_3'(\phi), y_3'(\phi)]$, then $O_-(q)$ is bounded

(ii) if $q \in S(x_1(\phi), y_1(\phi)) \supset S[x_3'(\phi), y_3'(\phi)]$,

then $J^+(q, \phi) \cap S = \emptyset$ by (6.7)

In either case, $q \notin \Omega(\phi) \setminus Per(\phi)$. $\square$

We codify (7.7) into a global statement by noting that for any compact region $C \subset \mathbb{R}^2$ there are finitely many transverse sections whose saturation contains all non-restpoints in C.

(7.8) *Proposition*:

(i) *Given* $C \subset \mathbb{R}^2$ *a compact region, there is a set* $\mathscr{G}_2(C) \subset \mathscr{G}_1(C)$ *of flows, dense-open in the* C^r *topology so that* $\psi \in \mathscr{G}_2(C)$ *implies* $C \cap \Omega(\phi) \subset Per(\phi)$. (ii) *Thus, there exists a* C^r*-residual set* $\mathscr{G}_2 \subset \mathscr{G}_1$ *consisting of flows such that*

$$\Omega(\phi) = Per(\phi).$$

For any $\phi \in \mathscr{G}_2$, *the separatrices of any saddle (fixed or at infinity) are distinct orbits.*

Proof:

(i) is immediate, (ii) is obtained by $\mathscr{G}_2 = \bigcap_i \mathscr{G}_2(C_i)$ for $C_i \subset C_{i+1}$, $\bigcup C_i = \mathbb{R}^2$. If $\phi \in \mathscr{G}_2$ and a saddle has a single orbit as both stable and unstable separatrix, then it is either a saddle connection or a non-periodic nonwandering orbit. $\square$

(7.9) *Remark*:

If $\phi \in \mathscr{G}_2(C)$ and p_i, $i=1,2,3$ form a triple with $p_1, p_3 \in C$, then $O(p_1, \phi) \neq O(p_3, \phi)$; because otherwise by (6.13) and (6.14) we can connect $O_-(p_1, \phi)$ with $O_+(p_2, \phi)$ to obtain flows near ϕ (hence in $\mathscr{G}_2(C)$) with unbounded non-wandering orbits. Thus, for $\phi \in \mathscr{G}_2$, a triple is always made up of three distinct ϕ-orbits. Finally, we have noted the distinction between two-sided and one-sided triples. Note that for a two-sided triple, $p_3 \in (J^+)^2(p_1) \setminus J^+(p_1)$, a situation which Klok [Kℓ] has shown to not exist generically. Thus we have

(7.10) *Proposition*

There is a C^r*-residual set* $\mathscr{G}_3 \subset \mathscr{G}_2$ *(r ≥ 1) such that any triple* p_i *i=1,2,3 for* $\phi \in \mathscr{G}_3$ *is one-sided, consists of three distinct orbits, and no transverse section intersects all three orbits.*

To complete the proof of theorem C, we need to show that generically there are no triples. We do this by means of a "local" lemma, which says that given three disjoint transverse sections S_i, i=1,2,3, for a generic flow no triple p_i, i=1,2,3 with $p_1 \in S_1$ and $p_3 \in S_3$ can have

$p_2 \in S_1 \cup S_2 \cup S_3$. We accomplish this "local" proof in several steps, the first of which is

(7.11) _Lemma_:

Given $\phi \in \mathcal{G}_3$ and two disjoint compact transverse sections S_1 and S_3, there exists a nonempty c^r-open set U with ϕ in its closure such that if $\psi \in U \cap \mathcal{G}_3$ and p_i, $i=1,2,3$ are a triple for ψ such that $p_1 \in S_1$ and $p_3 \in S_3$, then $0(p_2, \psi)$ does not intersect both S_1 and S_3.

Proof:

Suppose $\phi \in \mathcal{G}_3$, $p_j \in S_j$ for $j=1,3$, and p_i, $i=1,2,3$ form a triple, with $0(p_2, \phi)$ crossing both S_1 and S_3 - say, $p_2 \in S_1$ and $p_2' \in 0_+(p_2, \phi) \cap S_3$. (see fig. 7.2a)

Using (6.15) perturb ϕ into an open set U_1 such that for $\psi \in U_1$ there are points

$$x_1(\psi) < x_2(\psi) < x_3'(\psi)$$

in S_1 such that $0_+(x_i, \psi)$ $(i=1,2)$ escape to infinity, $0_+(x_1, \psi)$ without crossing S_3 but $0_+(x_2, \psi)$ crossing S_3 at $x_2'(\psi)$, and such that $z < x_3'(\psi)$, $z \in S_1$ implies $0_-(z, \psi)$ bounded. We can insure also that there is a point $r(\psi) \in S_3$, on the same side of

$$L_\psi = 0_+(x_1, \psi) \cup S_1[x_1, x_3'] \cup 0_+(x_3, \psi)$$

as x_2 (and p_3) so that $0_-(r(\psi), \psi)$ also escapes to infinity. (fig. 7.2b).

This eliminates triples crossing S_1 below x_3' or S_3 above x_2'. Suppose now that some $\psi \in U_1 \cap \mathcal{G}_3$ has a triple q_i, $i=1,2,3$ with $q_i \in S_i$ $(i=1,3)$ and $q_2 \in S_1$, $q_2' \in 0_+(q_2, \psi) \cap S_3$. We investigate in some detail the relative positions of q_1, q_2 in S_1 and q_2', q_3 in S_3. We have noted already that $q_2 > x_3'$, since $0_-(q_2, \psi)$ is unbounded. Similarly, since every unbounded semi orbit crossing $S_1[x_3', q_2]$ must later cross $S_3[q_2', x_2']$ and $0_+(q_1, \psi)$ cannot cross S_3, we must have $q_1 > q_2$, and similarly $q_3 < q_2'$ (fig. 7.2c)

Again applying (6.15) we perturb ψ into an open set $U_2 \subset U_1$ in which every vectorfield ξ has points $y_3'(\xi) < y_2(\xi) < y_1(\xi)$ in S_1 analogous to $x_i(\psi)$: $0_+(y_i, \xi)$ escape to infinity for $i=1,2$, $0_+(y_1)$ without hitting S_3 but $y_2' \in 0_+(y_2, \xi) \cap S_3$, and $z > y_3'$ in S_1 implies $0_-(z, \xi)$ bounded.

We claim that $\phi \in U_2 \cap \mathcal{G}_3$ cannot have a triple r_i, $i=1,2,3$ with r_i, $r_2 \in S_1$, r_2', $r_3 \in S_3$, $r_2' \in 0_+(r_2, \phi)$. Otherwise, since $0_-(r_2)$ is unbounded, $x_3' < r_2 < y_3'$. But $x_2 < x_3'$ and $y_2 > y_3'$ both have orbits which cross S_3 and then escape to infinity. Since r_2' must lie between x_2' and

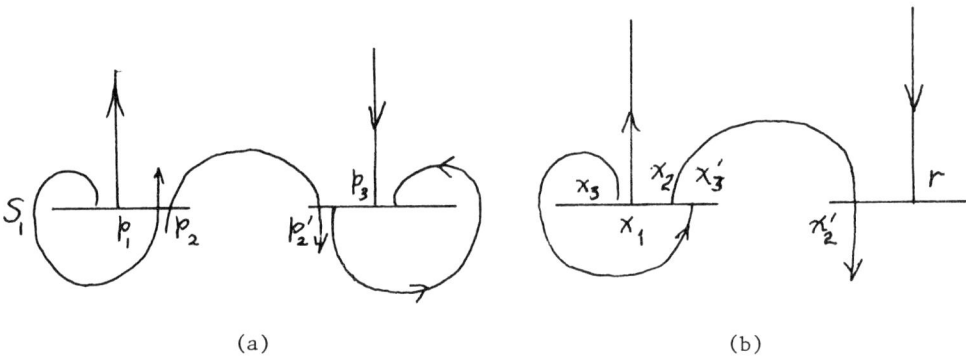

(a)   (b)

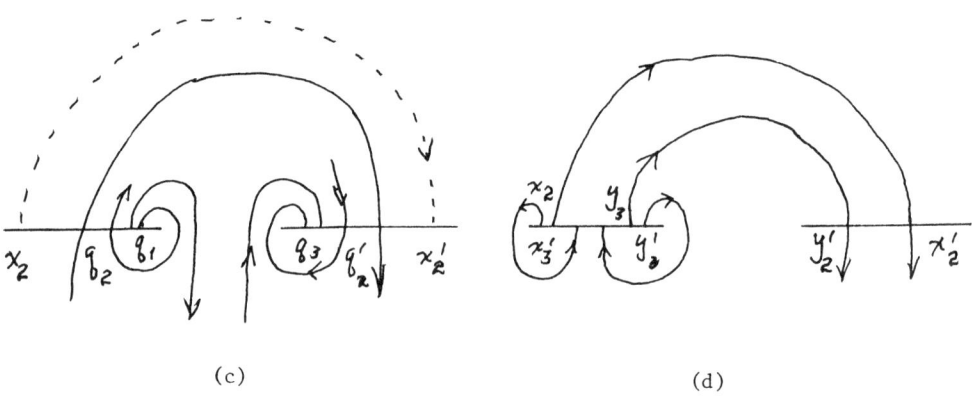

(c)   (d)

Figure 7.2

y_2' on S_3, it is impossible that $J^+(r_2', \phi) \cap S_3 \neq \emptyset$ (by 6.7)) and so r_i cannot form a triple. (see fig. 7.2d). $\square$

Having removed the possibility that the middle orbit crosses both of the other transversals, we now remove the further possibility that it crosses either one.

(7.12) Lemma:

Given S_1 and S_3 sections as in (7.11), let U be the open set given by the conclusion of (7.11). There exists a dense open subset $V \subset U$ such that if $\phi \in V \cap \mathcal{G}_3$ and p_i (i=1,2,3) form a triple with $p_i \in S_i$, i=1,3, then $O(p_2, \phi) \cap [S_1 \cup S_3] = \emptyset$.

Proof:

Suppose $\phi \in U \cap \mathcal{G}_3$ has a triple with $p_1, p_2 \in S_1$ and $p_3 \in S_3$, we can use (6.13-15) to perturb ϕ into an open set $V_1 \subset U$ such that $\psi \in V_1$ has points $x_1(\psi) < x_2(\psi) < x_3(\psi)$ in S_1 and $x_2'(\psi) < r(\psi)$ in S_3 such that

    (i) $x_2' \in O_+(x_2, \psi)$

    (ii) $O_+(x_1, \psi)$ and $O_+(x_3, \psi)$ escape to infinity without crossing S_3

    (iii) $O_-(x_2, \psi)$ escapes to infinity

    (iv) $O_-(r, \psi)$ escapes to infinity without crossing S_1.

(see fig. 7.3).

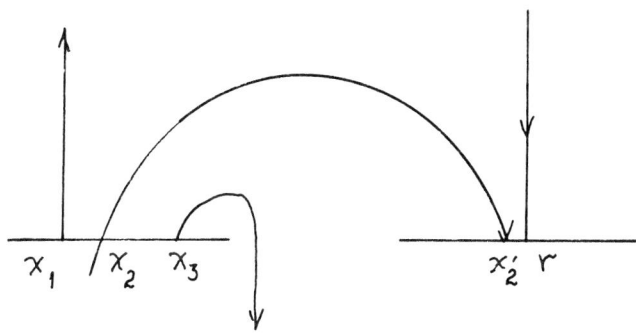

Figure 7.3

Now, if $\psi \in V_1 \cap \mathcal{G}_3$ has another triple q_i $(i=1,2,3)$ so that $q_i \in S_i$ $(i=1,3)$, we know $\mathcal{O}(q_2, \psi)$ cannot cross both S_1 and S_3; suppose $q_2 \in S_1$.

By (6.7), q_1 cannot belong to $S_1[x_1, x_3]$, in fact q_2 must separate q_1 from x_1 and x_3 in S_1. Since $\mathcal{O}(q_2, \psi)$ separates the plane and does not cross S_3, while $\mathcal{O}(x_2)$ does cross S_3, we see that q_1 and S_3 (hence q_3) lie on opposite sides of $\mathcal{O}(q_2, \psi)$, contradicting the fact that any triple is one-sided. Thus, for any $\psi \in V_1 \cap \mathcal{G}_3$ and any triple q_i $(i=1,2,3)$ with $q_1 \in S_1$, $q_3 \in S_3$, we have $\mathcal{O}(q_2, \psi) \cap S_1 = \emptyset$. A similar argument produces $V \subset V_1$ so that $\phi \in V \cap \mathcal{G}_3$ also implies $q_2 \notin S_3$ in the case above. $\square$

We have now generically removed the possibility that two out of the three orbits in a triple cross the same transversal. Now we (locally) eliminate triples altogether.

(7.13) *Lemma:*

Given three disjoint compact arcs S_i, *$i=1,2,3$ all transverse to* ϕ, *there exists a* C^r*-open set* $W \subset V$ *with* ϕ *in its closure such that for any* $\psi \in W \cap \mathcal{G}_3$ *and any triple* p_i *$(i=1,2,3)$ with* $p_i \in S_i$ *for* $i=1,3$, $\mathcal{O}(p_2, \psi)$ *does not cross* S_2.

Proof:

Suppose $p_i \in S_i$, $i=1,2,3$ form a triple for $\phi \in V \cap \mathcal{G}_3$. Using (6.14) perturb ϕ into an open set W_1 with ϕ in its closure so that $\psi \in W_1$ has points $x_i(\psi) \in S_i$, $i=1,2,3$ and $x_2'(\psi) \in S_3$ such that (fig. 7.4a)

(i) $\mathcal{O}_+(x_1, \psi)$ escapes to infinity without crossing S_2 or S_3

(ii) $\mathcal{O}_-(x_2, \psi)$ escapes to infinity without crossing S_1 or S_3

(iii) $\mathcal{O}_-(x_3, \psi)$ escapes to infinity without crossing S_1 or S_2.

The curve

$$L_1 = \mathcal{O}_-(x_2', \psi) \cup S_3[x_2', x_3] \cup \mathcal{O}_-(x_3', \psi)$$

separates $\mathbb{R}^2$ into two components $A_\pm^1$, with $S_1 \subset A_+^1$.

If $\psi \in W_1 \cap \mathcal{G}_3$ has a triple $q_i \in S_i$, $i=1,2,3$ then clearly the q_i all belong to A_+^1. Using (6.14) again, perturb further into an open set $W_2 \subset W_1$ such that every $\xi \in W_2$ has $y(\xi) \in S_1$ and $y'(\xi) \in \mathcal{O}_+(y, \xi) \cap S_2$, so that $\mathcal{O}_+(y', \xi)$ escapes to infinity without crossing S_1 or S_3. Again

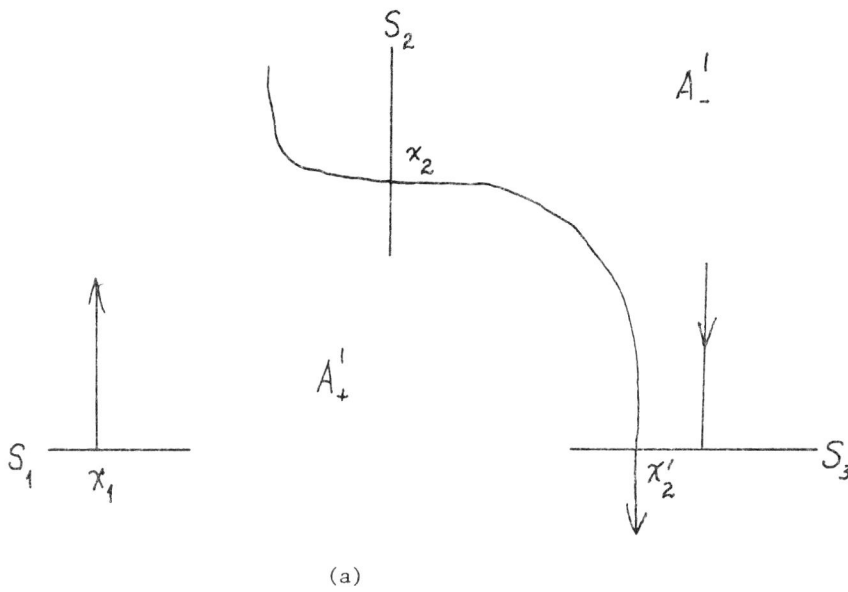

(a)

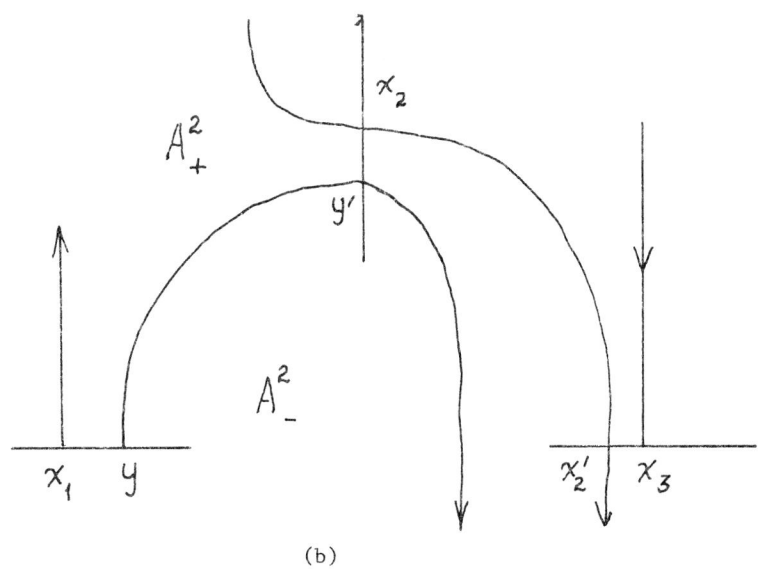

(b)

Figure 7.4

$$L_2 = 0_+(x_1, \xi) \cup S_1[x_1, y] \cup 0_+(y_1, \xi)$$

separates $\mathbb{R}^2$ into components $A_\pm^2$, with $S_3 \subset A_+^2$. (fig. 7.4b)

Finally, if $\eta \in W_2 \cap \mathcal{G}_3$ has a triple $r_i \in S_i$ then $r_2 \in A_+^1 \cap A_+^2$, and hence r_2 lies between x_2 and y' on S_2. On the other hand, $0(r_2) \cap [S_1 \cup S_3] = \emptyset$, and $0(r_2)$ separates $y \in S_1$ from $x_2' \in S_3$, contradicting the one-sidedness of triples. $\square$

These results let us formulate the "compact region" version of the generic absence of triples:

(7.14) *Proposition:*

Given a compact region $C \subset \mathbb{R}^2$*, there is a* C^r*-open dense set* $\mathcal{G}_4(C)$ *(r* $\geq$ *1) such that if* $\phi \in \mathcal{G}_4(C) \cap \mathcal{G}_3$ *and* p_i *(i=1,2,3) form a triple for* ϕ *with* p_1, $p_3 \in C$*, then* $0(p_2, \phi)$ *does not intersect* C.

Proof:

Given $\phi \in \mathcal{G}_3$, we find finitely many compact disjoint sections S_i such that any ϕ-orbit not a restpoint intersecting C crosses one of the S_i. Applying (7.11-7.13) to these sections in triples, we obtain an open set U_ϕ with ϕ in its closure for which no triple with p_1, $p_3 \in C$ (for a flow in $\mathcal{G}_3$) can have $p_2 \in S_i$ for any i. The union of all such sets U_ϕ for $\phi \in \mathcal{G}_3$ given a dense open set $\mathcal{G}_4(C)$. $\square$

Putting all of this together, we obtain theorem C, which we state in a somewhat elaborate form:

(7.15) *Theorem C:*

For $r \geq 1$*, there is a* C^r*-residual set* $\mathcal{G}_4$ *of flows on* $\mathbb{R}^2$ *such that any* $\phi \in \mathcal{G}_4$ *satisfies:*

(i) $\Omega(\phi) = Per(\phi)$ *consists of hyperbolic orbits*

(ii) *there are no oscillating orbits*

(iii) *there are are no connections between fixed saddles*

(iv) *the separatrices of any fixed saddle do not form a saddle at infinity with any semi-orbit*

(v) *there are no triples.*

We note that (iii)-(v) is simply the elaboration of $\overline{W}^-(\phi) \cap W^+(\phi) = \emptyset$.

Proof:

Write $\mathbb{R}^2 = \bigcup_{i=1}^{\infty} C_i$, where $C_i \subset C_{i+1}$ are compact regions, and set

$$\mathcal{G}_4 = \mathcal{G}_3 \cap \bigcap_{i=1}^{\infty} \mathcal{G}_4(C_i).$$

(Recall that $\mathcal{G}_1 \supset \mathcal{G}_2 \supset \mathcal{G}_3$ are described, respectively, in (7.6), (7.8), and (7.10).) Then $\mathcal{G}_4$ is residual (the Baire property is proved in [Pe 3], and (i) follows from (7.8), (ii)-(iv) from (7.6), and (v) from (7.14). $\quad\square$

8. *Necessary conditions for structural stability*
of flows in the plane

This section is devoted to the proof of theorem B, the necessity of (3.1) for global structural stability of a flow in the plane. At the end, we reconsider briefly the use of the C^4 topology in the example given by corollary (2.9).

We recall theorem B, stated in full:

(8.1) *Theorem B:*
   If a flow ϕ on $\mathbb{R}^2$ is globally C^r structurally stable $(r \geq 1)$ then:
   (i)   *there are no nontrivial minimal sets and no*
         oscillating orbits
   (ii) *every orbit in $Per(\phi)$ is hyperbolic*
   (iii) *$clos\ W^-(\phi) \cap clos\ W^+(\phi) \subset Per(\phi)$*
   (iv) *$\Omega(\phi) = Per(\phi)$.*

Note that conditions (i), (iii) and (iv) are invariants of topological equivalence, while (ii) involves the derivative. It is clear that any condition which is invariant under topological equivalence and is true for a C^r-dense set of flows must hold for any globally C^r structurally stable flow, since every stable flow is equivalent to some element of each dense set. Thus, an immediate corollary of theorem C is:

(8.2) *Proposition:*
   If ϕ on $\mathbb{R}^2$ is globally C^r structurally stable $(r \geq 1)$, then
   (i)   *there are no nontrivial minimal sets and no*
         oscillating orbits
   (ii) *every periodic orbit is a limit cycle (an*
         attractor or a repellor) and every restpoint
         is a fixed attractor or repellor or is a
         topological saddle.
   (iii) *$W^-(\phi) \cap W^+(\phi) = \emptyset$*
   (iv) *$\Omega(\phi) = Per(\phi)$.*

(8.2a) *Remarks:*
   We note that by (6.4) or (6.5) no flow on $\mathbb{R}^2$ can have non-trivial minimal sets. Also, (ii)-(iv) imply there are no oscillating orbits. In (ii), by an attractor (repellor) we mean an orbit γ with a neighborhood U such

that every positive (negative) semi-orbit entering U stays in U and tends toward γ. By a topological saddle we mean a fixedpoint p such that the flow on some neighborhood U of p is topologically equivalent to that near a hyperbolic fixed saddle.

There are two respects in which (8.2) needs to be improved. We need to show periodic orbits hyperbolic in (ii), and to consider the closures of the separatrix sets in (iii).

We have already seen in §2 (figure 2.8) that the condition (iii) in (8.1) is not generic. However, the necessity of (8.1(iii)) follows from (8.2(iii)), because failure of the first at ϕ implies failure of the second nearby.

(8.3) _Lemma_:

   If ϕ is a flow on the open surface M for which $\Omega(\phi) = Per(\phi)$ consists of topologically hyperbolic orbits, then if $clos\ W^-(\phi)$ intersects $clos\ W^+(\phi)$ at a non-saddle x, there exist C^r perturbations ψ of ϕ for all $r \geq 1$ such that $W^-(\psi) \cap W^+(\psi) \neq \emptyset$.

Proof:

   Since there are no sinks (resp. sources) in $clos\ W^+(\phi)$ (resp. $clos\ W^-(\phi)$), x cannot be periodic, hence wanders. Let F be a flowbox neighborhood of x; by making it wandering, we can assume $O_\pm(S_\mp, \phi)$ does not intersect F. ($S_\mp$ are the exit and entrance sets of F.) Set $x_\mp = O_\pm(x, \phi) \cap S_\mp$.

   By assumption, there exist $x_n^\pm \to x_\pm$ in $S_\pm$ such that $x_n^\pm \in W^\pm(\phi)$. Thus, for each n, $O_+(x_n^-)$ $(O_-(x_n^+))$ is the stable (unstable) separatrix of a fixed saddle p_n (q_n) or forms a saddle at infinity with some $O_-(y_n)$ $(O_+(z_n))$, which we can assume disjoint from F. Pick a C^r neighborhood U of ϕ, and find a flowbox neighborhood $\widetilde{F} \subset F$ of x as in (6.12). For large n, $x_n^\pm \in S_\pm$; thus there exists $\psi \in U$ with $\dot\psi = \dot\phi$ off F such that $x_n^- \in O_+(x_n^+, \psi)$. Since $O_\pm(x_n^\mp, \phi) = O_\pm(x_n^\mp, \psi)$, this gives

$$x_n^- \in W^-(\psi) \cap W^+(\psi),$$

as desired. $\square$

(8.4) _Corollary_:

   If ϕ is a globally C^r structurally stable flow on $\mathbb{R}^2$, then $clos\ W^-(\phi) \cap clos\ W^+(\phi)$ consists of fixed saddles.

Proof:

   Otherwise, (8.3) gives $W^-(\psi) \cap W^+(\psi) \neq \emptyset$ for some ψ near ϕ, hence (since ψ and ϕ are equivalent) for $\psi = \phi$, contradicting (8.2iii). $\square$

The proof that structural stability implies hyperbolicity of orbits in Per(ϕ) is fairly straightforward, but for technical reasons we precede it with a few perturbation lemmas. The basic argument here is implicit in [M 2,4] and is also given in [Kr 5].

(8.5) Lemma:

Given $f : \mathbb{R}^n \to \mathbb{R}^n$ a C^r map, let $\widetilde{f}$ denote the rth Taylor polynomial of f at $0 : \widetilde{f}(x) = f(0) + Df(0)x + \ldots + D^r f(0)(r, \ldots, r)$. For every strong C^r neighborhood U of f and every neighborhood V of 0 , there exist $g : \mathbb{R}^n \to \mathbb{R}^n$ and a neighborhood U of 0 , clos $U \subset V$, such that

(i) $g \in U$

(ii) $g(x) = f(x)$ off V

(iii) $g(x) = \widetilde{f}(x)$ on U .

Proof:

Fix a C^∞ bump function $h : \mathbb{R}^n \to [0,1]$ such that

$$h(x) = 1 \quad \text{if} \quad \|x\| \leq 1/2$$
$$h(x) = 0 \quad \text{if} \quad \|x\| \leq 1$$

Note that for $\|x\| > 1$ (and also for $\|x\| < 1/2$) $h(x)$ is constant, so that every derivative vanishes.

Given $a > 0$, define a function $f_a(x)$ by

$$f_a(x) = [1 - h(ax)] \, f(x) + h(ax) \, \widetilde{f}(x) .$$

Then

$$f(x) - f_a(x) = h(ax) \, [f(x) - \widetilde{f}(x)]$$

and Leibniz's rule gives the derivatives of this as

$$D^n f(x) - D^n f_a(x) = \sum_{k=0}^{n} \binom{n}{k} a^k [D^k h(ax)][D^{n-k} f(x) - D^{n-k}\widetilde{f}(x)] .$$

Note that in this formula, every term except $k=n$ vanishes when $\|x\| > 1/a$. Furthermore, Taylor's formula gives us the estimate

$$\| D^m f(x) - D^m \widetilde{f}(x) \| < \varepsilon_m(x) \|x\|^{r-m}$$

where $\varepsilon_m(x) \to 0$ as $\|x\| \to 0$.

Thus, we have an estimate on the C^r -distance between f and f_a as follows

$$\sup_{x \in \mathbb{R}^n} \| D^n f(x) - D^n f_a(x) \|$$

$$\leq \sum_{k=0}^{n} \binom{n}{k} a^k \sup_{x \in \mathbb{R}^n} \| D^k h(x) \| \sup_{\|x\| < 1/a} \| D^{n-k} f(x) - D^{n-k} \widetilde{f}(x) \|$$

and setting $\beta_k = \sup_{\mathbb{R}^n} \| D^k h(x) \|$,

we have for $n \leq r$

$$\sup_{x \in \mathbb{R}^n} \| D^n f(x) - D^n f_a(x) \|$$

$$\leq \sum_{k=0}^{n} \binom{n}{k} a^k \beta_k a^{n-k-r} \sup_{\|x\| < 1/a} \| \varepsilon_{n-k}(x) \|.$$

Note that as $a \to \infty$, $1/a \to 0$ and this last sum goes to zero uniformly. Thus, given V and U, we can make a large enough that $f_a \varepsilon\ U$ and that $U_1 = \{x \mid \|x\| \leq 1/a\}$ and hence $U = \{x \mid \|x\| \leq 1/2a\}$ are contained in V. Since $h(ax) = 0$ for $\|x\| > 1/a$ and 1 for $\|x\| < 1/2a$, we see that $g = f_a$ satisfies (i), (ii), and (iii). $\square$

(8.6) Corollary:

Given f, $\widetilde{f}$, U, U *and* V *as above, there exists a neighborhood* V *of* $\widetilde{f}$ *in the space of polynomials of degree* r *such that given* $g \varepsilon V$, *there exists* $g \varepsilon U$ *with* $g(x) = \widetilde{g}(x)$ *on* U.

Proof:

It suffices to prove the lemma with $f = f_a$ as above. But given U and $k(x) = h(ax)$ as above, note that the function $\varepsilon k(x)$ has C^r size going to zero uniformly as $\varepsilon \to 0$; furthermore, the coefficent topology on polynomials of degree r agrees with the (compact-open) uniform C^r topology concentrated on V. Thus,

$$g(x) = f_a(x) + k(x)\, p(x)$$

where p is a polynomial of degree r, satisfies

$$\| g - f_a \|_{C^r} \leq \| k(x) \|_{C^r} \| p(x) \|_r$$

where the estimate on the C^r-size of $p(x)$ is concentrated on the support of k, which is a subset of V. Thus, we pick a bound on the size of V so that $\widetilde{f} + p \varepsilon V$ implies $\|k\|_r \|p\|_r$ small, and thus $g \varepsilon U$. $\square$

(8.7) Corollary:

  (i) *given a fixedpoint* x *of* ϕ, *a neighborhood* V
    of x, *and a* C^r *neighborhood* U *of* ϕ, *there*
    exist neighborhoods $U \subset V$ *of* x *and* V *of the*
    r^{th} *Taylor polynomial* X *of* $\dot{\phi}$ *at* x *such that*
    for each $Y \varepsilon V$ *there exists* $\psi \varepsilon U$ *with*

    (a) $\dot{\psi} = \dot{\phi}$ *off* V

    (b) $\dot{\psi} = Y$ *on* V.

(ii)   given a periodic orbit γ of ϕ , let S be a transverse
section through $x \in \gamma$ and $P_\phi : S \to S$ the Poincaré
map of ϕ . Given a c^r neighborhood U of ϕ and a
neighborhood U of γ there exist c^r neighborhoods
V of P_ϕ and a neighborhood $S' \subset S$ of x in S
such that for any $\pi \in V$ with $\pi(x) = x$ there exists
$\psi \in U$ such that $\dot\psi = \dot\phi$ off U , and $P_\psi(y) = \pi(y)$
for $y \in S'$.

Proof:

For (a), apply (8.6) to $\dot\phi = f$.

For (b), pick a flowbox neighborhood $F \subset U$ of x .
It suffices to show that any Poincaré map $\widetilde\pi : S_+ \to S_-$ near $\widetilde P_\phi : S_+ \to S_-$
can be realized on some $S' \subset S_+$ via a c^r perturbation ψ of ϕ on F .
But this is quite easy to do by noting that if $\widetilde\pi$ and $\widetilde P_\phi$ are near each
other as diffeomorphisms of the interval, there is an isotopy between them
which is always near $\widetilde P_\phi$. This can be realized via a time-dependent vectorfield, which we can embed in a flowbox $\widetilde F \subset \text{int } F$ using bump functions . $\square$

With these perturbation lemmas, we have the opportunity to "overkill"
the rest of theorem C :

(8.8)   Proposition:

If ϕ is a globally c^r structurally stable flow on $\mathbb{R}^2$ (or any open
surface) $(r \geq 1)$ then every orbit in $Per(\phi)$ is hyperbolic .

Proof:

We adapt the counting argument in [M 2,4] and [R 1]. From (8.2) we know
that any orbit γ in $Per(\phi)$ is equivalent to a hyperbolic one, hence has
a neighborhood V which contains no other orbits in $Per(\phi)$. By nesting
two such neighborhoods, one inside the other, we can find a compact-open
neighborhood N of the identity such that any flow ψ equivalent to ϕ via
$h \in N$ has a unique orbit $\gamma(\psi)$ in $Per(\psi)$ contained in V . Pick U a
c^r neighborhood of ϕ as in the definition of structural stability so
$h \in N$ if $\psi \in U$.

We examine the stability type of various $\psi \in U$. Since $\gamma(\phi)$ is not
hyperbolic we have a fixedpoint with either a pair of complex eigenvalues
$\pm i \theta$ or at least one zero eigenvalue, or a periodic orbit which is a fixedpoint x of its Poincaré map $P_\phi : S \to S$ with $DP_\phi(x) = 1$. Using (8.7),
there exist perturbations of the Taylor polynomial of $\dot\phi$ (or P_ϕ) whose
linear parts are hyperbolic and possess eigenvalues $\alpha \pm i \theta$, α , or $1 + \alpha$
for all $\alpha \in (-\epsilon , \epsilon)$ for some $\epsilon > 0$ given. Now the cases with $\alpha > 0$ and

those with $\alpha < 0$ are not conjugate locally, and this contradicts the stability of ϕ, since both occur in U. $\square$

We close this section with a brief reconsideration of the "unencumbered"
notion of structural stability discussed in §2 (where no limitation is put on
h). Recall that in (2.9) we showed that C^4 "unencumbered" stability does
not imply hyperbolicity of orbits in Per(ϕ) (specifically, of fixedpoints).
On the other hand, in the C^1 topology, it does. The idea is that instead
of counting, we produce behavior not conjugate to any hyperbolic behavior.
Similar ideas appear in [F 1] and [Kr 5].

(8.9) Proposition:

*Suppose the flow ϕ on $\mathbb{R}^2$ is C^1 structurally stable in the sense
that it is topologically equivalent (via some h, not necessarily near the
identity) to any strong C^1 perturbation. Then ϕ is globally C^1 structurally stable.*

Proof:

We show that ϕ fulfills the hypotheses of theorem A. Noting that
(8.2-4) never use the distance between h and id, we need only prove that
all orbits in Per(ϕ) are hyperbolic.

Note that a linear vectorfield which is not hyperbolic has a restpoint
which fails to be isolated in Per(ϕ): either it is surrounded by infinitely
many circular orbits, or it sits on a line of restpoints. Similarly, a
periodic orbit whose Poincaré map is linear and has a non-hyperbolic fixedpoint is part of a band of periodic orbits.

(8.7) tells us that ϕ can be C^1-approximated by a flow ψ which, near
a given orbit $\gamma \in$ Per(ϕ), has $\gamma \in$ Per(ψ) and $\dot{\psi}$ (for restpoints) or P_ψ
(for circular orbits) linear near ψ, and equal to the linearization of $\dot{\phi}$
(or P_ϕ). Thus, if the latter is non-hyperbolic, we obtain one of the
phenomena of the previous paragraph. But none of these are conjugate to the
behavior near hyperbolic orbits, contradicting (8.2ii). $\square$

9. *Related questions*

We close with a brief discussion of several questions related to the concerns of this paper.

We already mentioned in section 1 our belief that theorem A gives necessary as well as sufficient conditions for structural stability on any surface. Of the three conditions, (3.1)(ii) is known necessary for global C^r structural stability on any surface, by (8.8). Any flow on any surface can be perturbed to eliminate non-trivial minimal sets : first, perturb to a C^2 flow, and note that any minimal set is contained in some compact region C. Then apply [Sch] to the flow on a closed manifold containing C (and perhaps some fixedpoints outside) to conclude that (if there is a non-trivial minimal set) the original manifold was the torus, and the flow was quasi-periodic. But these flows are easily perturbed into periodic ones.

This leaves two conditions : nonexistence of oscillating orbits, and condition (3.1)(iii). We believe that oscillating orbits are impossible in a structurally stable flow (on a surface), but formulate a stronger question.

(9.1) Problem:

On any open surface M , are the flows with no oscillating orbits C^r-generic for $r \geq 1$?

The other condition of theorem A, that clos $W^\pm(\phi)$ meet only in fixed saddles, is never generic, and on some surfaces even the weaker condition of theorem C (disjointness of $W^\pm(\phi)$) fails to be generic. Nevertheless, we believe the failure of either condition is an obstruction to structural stability:

(9.2) Conjecture:

On any open surface, any globally C^1 structurally stable flow ϕ must satisfy

$$W^-(\phi) \cap W^+(\phi) = \emptyset$$

and hence (by 8.3)

$$\text{clos } W^-(\phi) \cap \text{clos } W^+(\phi) \subset \text{Per}(\phi) .$$

Related to these perturbation questions is the extension to open surfaces of

(9.3) Theorem (Shub [Sh]):

Any dynamical system ϕ on a closed manifold M can be C^0-approximated

by a C^1-structurally stable system.

We believe

(9.3a) Conjecture:

Any flow ϕ on an open surface M can be strongly C^0-approximated by a globally C^1 structurally stable flow.

In particular, (9.3a) implies that every surface supports a structurally stable flow. This is known independently of (9.3a) - as noted in [Kr 1], every manifold supports a gradient flow without critical points, which by [Ni 1] is interior to the set of completely unstable flows. In [Kr 6] it is shown that every such flow on an open surface can be C^0-approximated by a C^1 flow with clos $W^-(\phi) \cap$ clos $W^+(\phi) = \emptyset$, and hence globally C^1 structurally stable. Furthermore, when M is an open surface with finite Euler characteristic, then [Ni 5] every completely unstable flow can be approximated by a non-explosive one, so for these manifolds (9.3a) holds when $\Omega(\phi) = \emptyset$.

Instead of generalizing theorems B and C to all open surfaces, one can return to the plane, and ask about special classes of flows. We find two such classes of interest : phase flows of second-order differential equations, and flows generated by polynomial vectorfields.

A second-order autonomous differential equation

(9.4a) $$\ddot{x} = f(x, \dot{x})$$

is represented in the (phase) plane by the vectorfield

(9.4b) $$\dot{x} = y$$
$$\dot{y} = f(x, y) .$$

In our theory, such a vectorfield can be multiplied by an nonvanishing real function, so that the vectorfields with the same integral curves as (9.4b) are characterized (up to multiplication by -1) by the condition

(9.4c) $$\dot{x} > 0 \quad \text{for} \quad y > 0$$
$$\dot{x} = 0 \quad \text{for} \quad y = 0$$
$$\dot{x} < 0 \quad \text{for} \quad y < 0 .$$

Note that replacing (9.4b) with some vectorfield satisfying (9.4c) amounts to a (phase-dependent) reparametrization of time in solutions of (9.4a).

One might ask about the occurence of the various phenomena we have considered for flows ϕ of vectorfields satisfying (9.4c), and their interpretation in terms of (9.4a). In the simplest case when Per(ϕ) is empty - or equivalently, when f(x, 0) is non-vanishing in (9.4a) - the flow is completely unstable, and one can ask (e.g. [M 1, p. 132, case 6]) about the

possible number of saddles at infinity and triples. Here, the completeness of
the *original* equation (9.4a) has a decided effect (remember that (9.4b) can
always be represented, *after reparametrization*, by a complete vectorfield
satisfying (9.4c)). When (9.4a) is complete, which is to say no solution
becomes unbounded or attains unbounded velocity in finite time, it is easy to
show that there is at most one saddle at infinity, (hence no triples) and its
stable (resp. unstable) separatrix is entirely contained in the upper (resp.
lower) half plane. Thus, when (9.4b) is completely unstable and complete, it
is topologically equivalent either to the parallelizable flow

(9.4d) $\dot{x} = y$

                                       $\dot{y} = 1$

(fig. 9.1a) or to the flow with one saddle at infinity

(9.4e) $\dot{x} = y$

                                       $\dot{y} = y^2 - 1$

(fig. 9.1b)

   On the other hand, when some solutions of (9.4b) have finite singularities, that is, they attain unbounded velocity but have a finite limiting
position, the situation is more complicated. If we take an even C^∞ "bump
function" h satisfying

$$h(y) = \quad 0 \quad \text{if} \quad |y| \geq 1$$
$$1 \quad \text{if} \quad |y| \leq 3/4$$
$$0 \leq h(y) \leq 1 \quad \text{for all} \quad y$$

and define an (incomplete) nonvanishing vectorfield by

(9.4f) $\dot{x} = y$

            $\dot{y} = (1 - y^2) h(y) + \dfrac{(1 - y^2)^2}{2} (\sin x) [h(y) - 1]$,

the integral curves are as follows : the horizontal lines $y = \pm 1$ are inte-
gral curves, and any integral curve outside the band $|y| \leq 1$ lies on the
graph of one of the parametrized family of functions

(9.4g) $y = \pm \left[\dfrac{k + 1 - \cos x}{k - \cos x} \right]^{1/2}$.

When $k > 1$, these graphs are curves lying above (or below) the band $|y| \leq 1$
and extending over $-\infty < x < +\infty$. However, for $|k| \leq 1$, this graph has
vertical asymptotes at $\cos x = k$. In particular, the graph for $k = 1$ has
vertical asymptotes at $x = 2n\pi$, $n \in \mathbb{Z}$ and minima at $(n\pi, \sqrt{3/2})$ for
$n \in \mathbb{Z}$ odd, and forms an infinite sequence of saddles at infinity for ϕ
(fig. 9.2a). Finally, if we take a second function $g(y) \geq 0$ such that

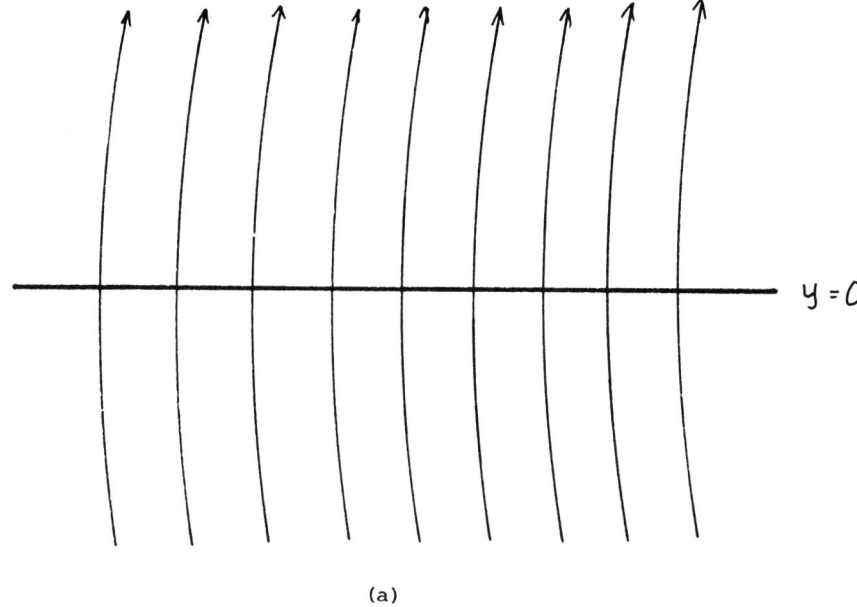

(a)

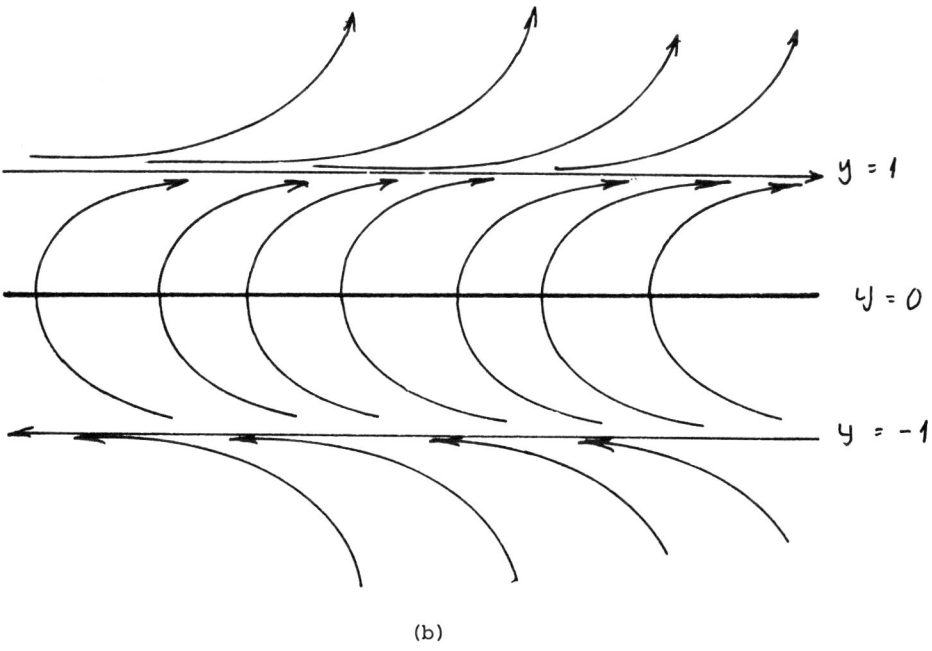

(b)

Figure 9.1

$$g(y) > 0 \quad \text{iff} \quad 6/5 < y < 7/5$$

and add $g(y)$ to the $\dot{y}$ term in (9.4f),

the unstable separatrix of each saddle excapes to infinity above the stable separatrix of the saddle to its right (fig. 9.2b), and we obtain a sequence of stable separatrices for saddles at infinity accumulating on some orbit similar to (9.4g) with $k > 1$. Thus, $W^+(\phi)$ is not closed in this case. In particular, using this phenomenon in place of fig. (2.7a) in a construction similar to fig. (2.7c), we can construct examples of (9.4c) for which clos $W^+(\phi)$ intersects a vertical section in an arbitrary closed set.

The dynamics of flows generated by a vectorfield

(9.5) $\dot{x} = P(x, y)$

                              $\dot{y} = Q(x, y)$

where P and Q are polynomials of a given degree r has received considerable attention, especially in connection with Hilbert's sixteenth problem. This concerns the kinds of dynamic phenomena possible for any flow (9.5). We wish to consider structural stability questions in this context. The set P_r of polynomial vectorfields on $\mathbb{R}^2$ of degree r or less is a discrete set in the strong C^r topology (in fact, in the strong C^0 topology) and so to perturb within P_r, one uses the *coefficient topology*, in which the coefficients of the various terms of P and Q are regarded as coordinates in euclidean space of dimension $r(r + 3)$. The coefficient topology and (uniform) C^r topology are the same for restrictions of polynomial vectorfields to any compact region.

Flows which are structurally stable in P_r (with coefficient topology) have been fully characterized by Tavares dos Santos [T], building on earlier work by Gonzalez Velasco [Go]. A flow ϕ on $\mathbb{R}^2$ can be represented via central projection (after reparametrization) as a flow ϕ on $S^2 \setminus S^1$, which (for polynomial fields) extends to the "circle at infinity" S^1. Then, [T] a flow is structurally stable in P_r if and only if every orbit in $\text{Per}(\phi)$ is hyperbolic and all saddle connections lie in S^1. Note that such a flow has finitely many orbits in $\text{Per}(\phi) = \Omega(\phi)$, and it is easy to check that the conditions of [T] imply (3.1iii). Thus, by theorem A, we have the positive part of the following

(9.6) _Remark:_

If a flow is structurally stable in P_r, _then it is also globally_ C^1 _structurally stable among all flows on_ $\mathbb{R}^2$. _The converse is false._

The falsity of the converse is shown by Chicone and Shafer [ChS]. They

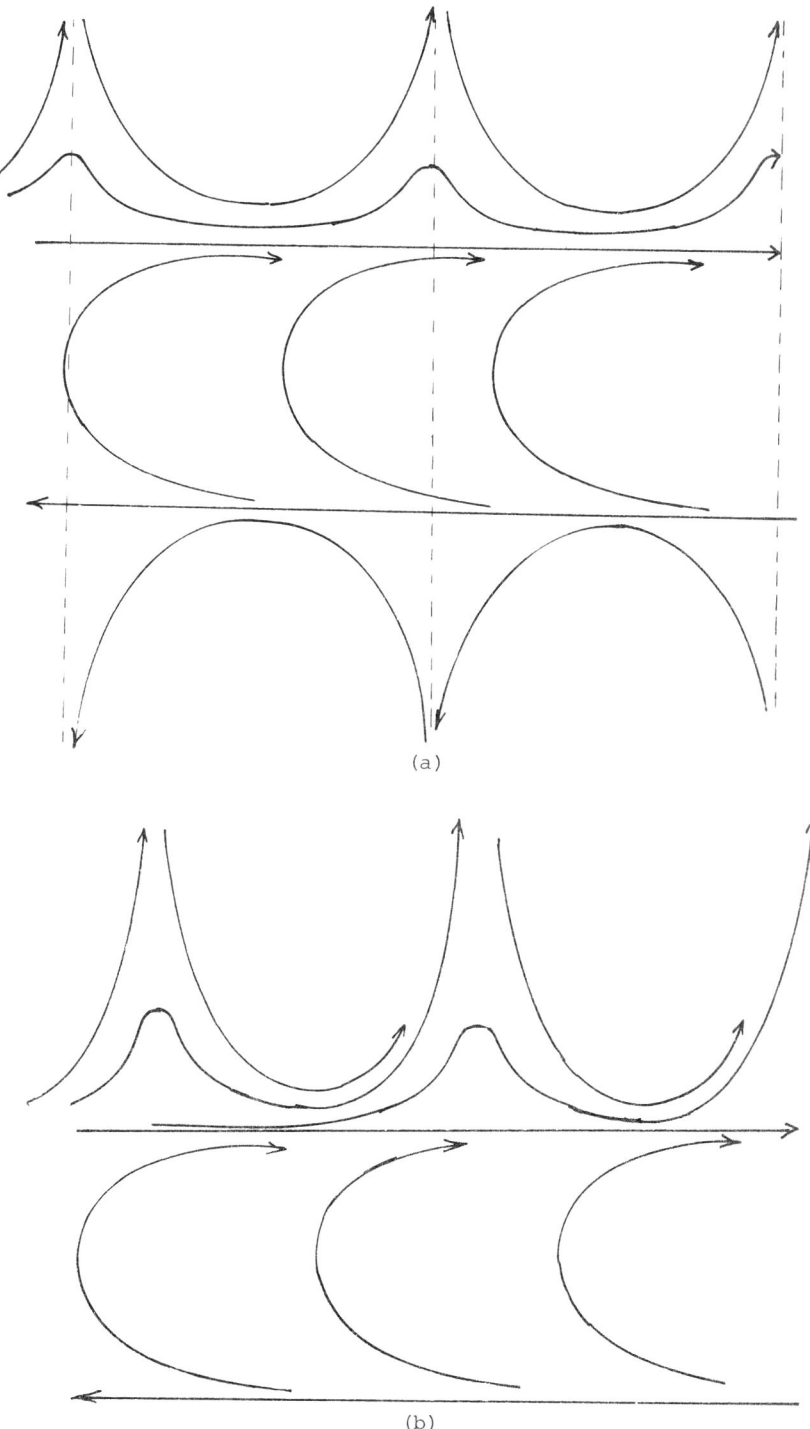

(a)

(b)

Figure 9.2

prove that a quadratic gradient vectorfield with $\text{Per}(\phi)$ finite and hyperbolic and no saddle connections also has $W^{\pm}(\phi)$ disjoint finite sets of orbits, and hence is globally C^1 structurally stable. On the other hand, the gradient of $H(x, y) = y(y^2 - x)$, given by

$$(9.6a) \qquad\qquad \begin{aligned} \dot{y} &= -y \\ \dot{y} &= 3y^2 - x \end{aligned}$$

has a fixedpoint at the origin (and no other finite fixedpoints) which is a hyperbolic saddle whose separatrices escape to infinity, but its central projection ϕ has a degenerate fixedpoint on S^1 which under perturbation in P_2 breaks into several restpoints, one of them finite. In particular, the quadratic gradient

$$(9.6b) \qquad\qquad \begin{aligned} \dot{x} &= -y + \varepsilon y^2 - 2\varepsilon x \\ \dot{y} &= 3y^2 - x + 2\varepsilon xy \end{aligned}$$

which is near (9.6a) in P_2 for ε small, has fixedpoints at the origin and at $(\varepsilon^{-2}, -\varepsilon^{-1})$.

Our final set of questions is the most important, but also the one which we are farthest from answering. This is the extension of structural stability theory to flows on open manifolds of dimension greater than two (and, as a special case, to diffeomorphisms on open surfaces). In this connection, it seems to us most important to achieve a thorough understanding of the kinds of prolongational relations involving unbounded orbits that persist under strong C^1 perturbation. One fundamental question of this type is

(9.7) *Problem* (Unbounded Closing Lemma):

   Suppose a dynamical system ϕ *has an unbounded nonwandering orbit*

$$0(x, \phi) \subset \Omega(\phi).$$

Are there dynamical systems ψ *arbitrarily near* ϕ *in the strong* C^1 *topology such that*

$$x \in \text{Per}(\psi) ?$$

This is not even answered for ϕ a flow on an arbitrary open surface.

A second aspect is invariant manifold theory. We state this less formally: suppose $J^+(x, \phi)$ contains a submanifold (which we tend to think of as unbounded). Then are there general conditions on ϕ of a "normal hyperbolicity" type which insure that for all ψ near ϕ (in the strong C^1 topology) some point $x(\psi)$ near x has $J^+(x(\psi), \psi)$ a submanifold? In fact, the notion of hyperbolicity for unbounded sets is not well understood. We have mentioned Osipov's unbounded version [O 1,2] of the Anosov structural stability theorem, which relies on uniform estimates in a preferred metric,

and White's example [Wh] of a complete metric on $\mathbb{R}^2$ under which
$(x, y) \to (x+1, y)$ has a hyperbolic structure. Thus, it seems that hyperbolicity is a very metric-dependent property, and may prove quite inappro-
priate for studying unbounded behavior. On the other hand, it would be useful to understand (i) given a metric under which an unbounded set Λ has
hyperbolic structure, which other metrics make Λ hyperbolic, and (ii) are
there metric-independent consequences of the existence of hyperbolic structure
in some metric (see [Me 2] for such a result on Anosov diffeomorphisms of
$\mathbb{R}^2$).

Closely tied to this is the understanding of the generic structure of
oscillating orbits. It is clear that (7.4) is false in higher dimensions.
We mention here a result of Wojtkowski [Wo], to the effect that certain open
surfaces have an open set of complete Riemannian metrics for which every point
lies on an uncountable number of oscillating geodesics. Another example, this
one for diffeomorphisms of $\mathbb{R}^2$ (and by suspension, flows on $D^2 x S^1$, where
D^2 in an open disc), illustrates the complicated nature of all the questions
above.

Start with a diffeomorphism $f : \mathbb{R}^2 \to \mathbb{R}^2$ with the following properties
(fig. 9.3)

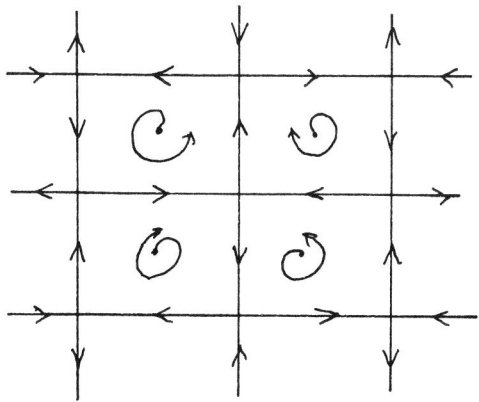

Figure 9.3

(i)   every point (m, n) ε ℤ × ℤ is a fixed
       saddle

(ii)   every point (p/2, q/2) , p, q odd is a
       fixed sink or source

(iii) if m an n have the same parity,
       $$W^+(m, n) = \{(x, n) \mid |x - m| < 1\}$$
       $$W^-(m, n) = \{(m, y) \mid |y - n| < 1\}$$
       while if they have opposite parity (one even,
       one odd) the roles of W^+ and W^- are re-
       versed

(iv)   each of the squares with corners on the integer
       lattice is invariant, with orbits tending to-
       ward the edge in one direction and toward the
       fixedpoint specified in (ii) in the other.

Now, consider the effect on f of a perturbation confined to a small
disk near some point halfway between the adjacent saddles, on their common
separatrix, which makes the unstable manifold of one saddle meet the stable
manifold of the other transversally, say $W^-(m, n) \pitchfork W^+(m+1, n)$.   Then
standard theory [Pa 2] tells us that for the perturbed diffeomorphism,
clos $W^-(m, n)$ (resp. clos $W^+(m+1, n)$) will include $W^+(m+1, n)$ (resp.
$W^-(m, n)$) .   If we do a similar perturbation on each edge of one square,
(fig. 9.4) the orbits homoclinic to one corner will contain a hyperbolic
cantor set with a dense orbit including points on every edge of the square.
If one follows this argument further, it seems reasonable that one can, by

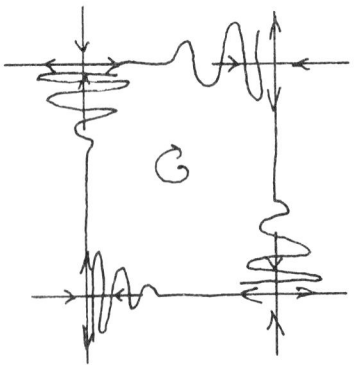

Figure 9.4

perturbing more and more saddle connections, create larger and larger hyperbolic sets, intersecting more and more squares, which have dense orbits. In the limit, we expect

(9.8) Conjecture:

There is a strong (resp. uniform) C^1 perturbation g of f with the same fixedpoint set as f such that the stable and unstable manifolds of any two fixed saddles have a point of transversal intersection. The set of homoclinic points to any fixed saddle of g contains an invariant, unbounded closed, perfect, totally disconnected set which has a dense orbit and a hyperbolic splitting (resp. with uniform estimates).

Note that if all the fixedpoints in (ii) are sources, then it seems likely that, after the appropriate perturbation, most positive semi orbits will be unbounded, and probably oscillating.

Of course, other examples on certain manifolds with similarly complicated behavior at infinity can easily be constructed by removing a fixedpoint or periodic orbit from complicated examples on closed manifolds (e.g., the horseshoe, Anosov diffeomorphisms, DA examples, etc.).

The example constructed above, however, has a far more irregular behavior at infinity than the ones obtained from closed manifolds. To handle both kinds of example at once, it is necessary to develop an intrinsic theory of these phenomena, not relying on a compactification of the original open manifold.

Institute of Mathematics
Warsaw Technical University

Institute of Mathematics
University of Warsaw

Department of Mathematics
Tufts University

ALGM   A. A. Andronov, E. A. Leontovich, I. I. Gordon, and A. G. Maier,
        Theory of Bifurcations of Dynamical Systems on a Plane. (Russian:
        Izd. Nanka, Moscow, 1967) = (English: Wiley, New York, 1973).

An   D. Anosov, Geodesic flows on closed Riemannian manifolds of negative
        curvature, *Proc. Steklov Inst. 90*(1967).

AP   A. A. Andronov and L. Pontrjagin, Systemes grossiers, *Comptes Rend.
        (Doklady) Acad. Sci. URSS 14*(1937) 247-250

Au1   J. Auslander, Some notions of stability and recurrence in dynamical
        systems, *Symp. System Thy. Poly. Inst. Brooklyn* (April, 1965) 115-120.

Au2   _____, Generalized recurrence in dynamical systems, *Contr. Diff.
        Eqns. III* (1964) 65-74.

AuS   J. Auslander and P. Seibert, Prolongations and stability in dynamical
        systems, *Ann. Inst. Fourier (Grenoble) 14*(1964) 237-267.

BhS   N. P. Bhatia and G. P. Szegö, *Stability Theory of Dynamical Systems*
        (Springer, NY, 1970).

Bo   R. Bowen, *On Axiom A Diffeomorphisms*, CBMS Reg. Conf. Series in Math.
        35 (AMS, 1978).

C   J. Collins, *Structural stability of completely unstable flows in the
        plane,* Ph.D. Dissertation, Tufts Univ., 1977.

Ch   C. C. Chicone, Quadratic gradients on the plane are generically Morse-
        Smale, *J. Diff. Eqns. 33*(1979) 159-166.

ChS   C. Chicone and D. Shafer, Quadratic Morse-Smale vector fields which
        are not structurally stable. Preprint, Columbia, Mo., 1981.

Co   C. Conley, *Isolated Invariant Sets and the Morse Index*, CBMS Reg.
        Conf. Series in Math. 38(1978).

Cp   W. A. Coppel, A survey of quadratic systems, *J. Diff. Eqns. 2*(1966)
        293-304.

DeB   H. F. DeBaggis, Dynamical systems with stable structures, *Contr. Thy.
        Non-Lin. Osc. II. Ann. Math. Study 29*(1952) 37-59.

DeM   W. C. deMelo, Structural stability of diffeomorphisms on two-mani-
        folds, *Invent. math. 21*(1973) 233-246.

F1   J. Franks, Necessary conditions for stability of diffeomorphisms,
        Trans. AMS 158(1971) 301-308.

F2   _____, Differentiably Ω-stable diffeomorphisms, *Topology 11*(1972)
        107-113.

F3   _____, Time dependent stable diffeomorphisms, *Invent. math. 24*(1974) 163-172.

G   J. Guckenheimer, Absolutely Ω-stable diffeomorphisms, *Topology 11*(1972) 195-197.

Go   E. A. González Velasco, Generic properties of polynomial vector fields at infinity, *Trans. AMS 143*(1969) 201-222.

Gu1   C. Gutierrez, Structural stability for flows on the torus with cross-cap, *Trans. AMS 241*(1978) 311-320.

Gu2   _____, Smooth nonorientable nontrivial recurrence on two-manifolds, *J. Diff. Eqns. 29*(1978) 388-395.

Gu3   _____, On two-dimensional recurrence, *Bol. Soc. Bras. Mat. 10*(1979) 1-16.

H   P. Hartman, *Ordinary Differential Equations*, (Wiley, 1964; re-issued by the author, 1973).

HP   M. W. Hirsch and C. C. Pugh, Stable manifolds and hyperbolic sets, *Proc. Symp. Pure Math. 14*(1970) 133-164.

I   M. Irwin, *Smooth Dynamical Systems* (Academic Press, 1980).

Ka   W. Kaplan, Regular curve-families filling the plane, I, II, *Duke Math. J. 7*(1940) 154-185, *8*(1941) 11-46.

Kℓ   F. Klok, Ω-stability of plane vectorfields, preprint, Groningen 1979.

Ko   J. Kotus, The vectorfields on $\mathbb{R}^2$ without oscillations are generic, *Demonstratio math.*, to appear.

Kr1   M. Krych, Note on structural stability of open manifolds, *Bull. Acad. Polon. Sci. 22*(1974) 1033-1038.

Kr2   _____, A generic property of non-vanishing vectorfields on $\mathbb{R}^2$, *Bull. Acad. Polon. Sci. 25*(1977) 361-368.

Kr3   _____, C^3-structurally stable diffeomorphism of the plane which is not C^2-structurally stable, *ibid* 869-873.

Kr4   _____, Two remarks on structural stability of plane dynamical systems, *Astérisque 50*(1977) 197-204.

Kr5   _____, *On the structural stability of diffeomorphisms and flows on noncompact manifolds*, (Polish), Ph.D. Dissertation, Warsaw, 1975.

Kr6   _____, On C^o density of structurally stable flows with no nonwandering points. To appear.

Ku   I. Kupka, Contribution à la theorie des champs génériques, *Contr. Diff. Eqns. 2*(1963) 457-484, *3*(1964) 411-420.

L   S. T. Liao, On the stability conjecture, *Chin. Ann. Math. 1*(1980) 9-30.

Ml   L. Markus, Global structure of ordinary differential equations in the plane, *Trans. AMS 76*(1954) 127-148.

M2   _____, Structurally stable differential systems, *Ann. Math. 73*(1961) 1-19.

M3   _____, Parallel dynamical systems, *Topology 8*(1969) 47-57.

M4   _____, The behavior of the solutions of a differential system near a periodic solution, *Ann. Math. 72*(1960) 245-266.

Ma1   R. Mañe, Contributions to the stability conjecture, *Topology 17*(1978) 383-396.

Ma2   _____, The stability conjecture on two-dimensional manifolds, to appear.

Me1   P. Mendes, On stability of dynamical systems on open manifolds, *J. Diff. Eqns. 16*(1974) 144-167.

Me2   _____, On Anosov diffeomorphisms on the plane, *Proc. AMS 63*(1977) 231-235.

Mo   J. Moser, On a theorem of Anosov, *J. Diff. Eqns. 5*(1969) 411-440.

Na1   T. Nadzieja, A remark on vector fields on open manifolds, *Asterisque 50*(1977) 315-322.

Na2   _____, Flows on open manifolds with positively Lagrange stable trajectories, *J. Diff. Eqns.*, to appear.

Ne1   D. A. Neumann, Completely unstable flows on 2-manifolds, *Trans. AMS 225*(1977) 211-226.

Ne2   _____, Classification of continuous flows on 2-manifolds, *Proc. AMS 48*(1975) 73-81.

NeO   D. A. Neumann and T. O'Brien, Global structure of continuous flows on 2-manifolds, *J. Diff. Eqns. 22*(1976) 89-110.

Ni1   Z. Nitecki, Explosions in completely unstable flows, I. Preventing explosions; II. Some examples, *Trans. AMS 245*(1978) 43-61, 63-88.

Ni2   _____, Bifurcation from completely unstable flows on the cylinder, *Ann. NY Acad. Sci. 316*(1979) 86-107.

Ni3   _____, On the topology of the set of completely unstable flows, *Trans. AMS 252*(1979) 147-162.

Ni4   _____, A note on explosive flows, *Lect. Notes in Math. 819*(1980) 364-379.

Ni5   _____, Recurrent structure of completely unstable flows on surfaces of finite Euler characteristic, *Am. J. Math. 103*(1981) 143-180.

Ni6   _____, *Differentiable Dynamics,* (MIT Press, 1971).

Nr   R. Narsimhan, *Analysis on real and complex manifolds,* (Amsterdam, North-Holland, 1968).

NS   V. V. Nemytskiĭ and V. V. Stepanov, *Qualitative Theory of Differential Equations,* (Russian: Goztekhizdat, Moscow, 1949) = (English: Princeton Univ. Press, Princeton, 1960).

O1   Ju. S. Osipov, Structural stability of noncompact Anosov flows and hyperbolic notions in Kepler's problem, (Russian) *Dokl. Akad. Nauk. SSSR 230*(1976) = (English) *Soviet Math. Doklady 17*(1977) 1389-1393.

O2   _____, The Kepler problem and geodesic flows in spaces of constant curvature, *Celestial Mech. 16*(1977) 191-208.

Pa1   J. Palis, On Morse-Smale dynamical systems, *Topology 4*(1969) 385-404.

Pa2   _____, A note on Ω-stability, *Proc. Symp. Pure Math. 14*(1970) 221-222

PaS   J. Palis and S. Smale, Structural stability theorems, *Proc. Symp. Pure Math. 14*(1970) 223-231.

Pe1   M. M. Peixoto, On structural stability, *Ann. Math. 69*(1959) 199-222.

Pe2   _____, Structural stability on two-dimensional manifolds, *Topology 1*(1962) 101-120, *2*(1963) 179-180.

Pe3   _____, On an approximation theorem of Kupka and Smale, *J. Diff. Eqns. 3*(1966) 214-227.

PePe   M. C. Peixoto and M. M. Peixoto, Structural stability in the plane with enlarged boundary conditions, *An. Acad. Bras. Cien. 31*(1959) 135-160.

PePu   M. M. Peixoto and C. C. Pugh, Structurally stable systems on open manifolds are never dense, *Am. J. Math. 87*(1968) 423-430.

Per   P. Percell, Structural stability on manifolds with boundary, *Topology 12*(1973) 123-144.

Pℓ1   V. A. Pliss, The location of separatrices of periodic saddle-point motion of second-order differential equations, *Diff. Eqns. 7*(1971) 906-927.

Pℓ2   _____, A hypothesis due to Smale, *Diff. Eqns. 8*(1972) 203-214.

Pℓu   F. Pluvinage, Espaces des feuilles de certaines structures feuilletees planes, *Coll. Math. 18*(1967) 89-102.

PS   C. C. Pugh and M. Shub, The Ω-stability theorem for flows, *Invent, math. 11*(1970) 150-158.

Pu   C. C. Pugh, The closing lemma, *Am. J. Math. 89*(1967) 956-1009.

R1   R. C. Robinson, C^r structural stability implies Kupka-Smale, *Dynamical Systems* (Academic Press, NY 1973) 443-450.

R2   _____, Structural stability of C^1 diffeomorphisms, *J. Diff. Eqns. 22*(1976) 28-73.

R3   _____, Structural stability of vector fields, *Ann. Math. 99*(1974) 154-175.

R4   _____, Structural stability of C^1 flows, *Lect. Notes in Math. 468*
            (1975) 262-277.

R5   _____, Structural stability on manifolds with boundary, *J. Diff.
            Eqns. 37*(1980) 1-11.

R6   _____, Stability, measure zero, and dimension two implies hyperboli-
            city, preprint, Evanston, 1976.

R7   _____, Stability theorems and hyperbolicity in dynamical systems,
            Rocky Mtn. J. Math. 7(1977) 425-437.

Ro   J. Robbin, A structural stability theorem, *Ann. Math. 94*(1971) 447-493.

Sch   A. J. Schwartz, A generalization of a Poincaré-Bendixson theorem to
            closed two-dimensional manifolds, *Am. J. Math. 85*(1963) 453-458.

Sh1   M. Shub, Stabilité globale des systémes dynamiques, *Astérisque 56*(1978)

Sh2   _____, Structurally stable systems are dense, *Bull. AMS 78*(1972)
            817-818.

Sm1   S. Smale, On gradient dynamical systems, *Ann. Math. 74*(1961) 199-206.

Sm2   _____, Differentiable dynamical systems, *Bull. AMS 73*(1967) 747-187.

Sm3   _____, The Ω-stability theorem, *Proc. Symp. Pure Math. 14*(1970) 289-
            298.

Sm4   _____, Stable manifolds for differential equations and
            diffeomorphisms, *Ann. Scuola Normale Superiore Pisa 18*(1963) 97-116.

So1   J. Sotomayor, Generic one-parameter families of vector fields on two-
            dimensional manifolds, *Publ. Math. IHES 43*(1973) 5-46.

So2   _____, Structural stability in manifolds with boundary, *Global
            Analysis and Its Applications 3*(Inter. Atomic Energy Agency, Vienna
            1974) 167-176.

T   G. Tavares dos Santos, Classification of generic quadratic vector
            fields with no limit cycles, *Lect. Notes Math. 597* (Springer, 1977)
            605-640.

TW   F. Takens and W. White, Vectorfields with no nonwandering points, *Am.
            J. Math. 98*(1976) 415-425.

U   T. Ura, Sur le courant extérieur á une région invariante; prolonge-
            ments d'une caracteristique et l'ordre de stabilité, *Funkcial. Ekvac.
            2*(1959) 143-200.

Wh   W. White, An Anosov translation, *Dynamical Systems* (Academic Press, NY
            1973) 666-670.

Wo   M. Wojtkowski, Oscillating geodesics on 2-dimensional manifolds,
            Asterisque 51(1978) 443-456.

General instructions to authors for
PREPARING REPRODUCTION COPY FOR MEMOIRS

> For more detailed instructions send for AMS booklet, "A Guide for Authors of Memoirs."
> Write to Editorial Offices, American Mathematical Society, P. O. Box 6248,
> Providence, R. I. 02940.

MEMOIRS are printed by photo-offset from camera copy fully prepared by the author. This means that, except for a reduction in size of 20 to 30%, the finished book will look exactly like the copy submitted. Thus the author will want to use a good quality typewriter with a new, medium-inked black ribbon, and submit clean copy on the appropriate model paper.

Model Paper, provided at no cost by the AMS, is paper marked with blue lines that confine the copy to the appropriate size. Author should specify, when ordering, whether typewriter to be used has PICA-size (10 characters to the inch) or ELITE-size type (12 characters to the inch).

Line Spacing – For best appearance, and economy, a typewriter equipped with a half-space ratchet – 12 notches to the inch – should be used. (This may be purchased and attached at small cost.) Three notches make the desired spacing, which is equivalent to 1-1/2 ordinary single spaces. Where copy has a great many subscripts and superscripts, however, double spacing should be used.

Special Characters may be filled in carefully freehand, using dense black ink, or INSTANT ("rub-on") LETTERING may be used. AMS has a sheet of several hundred most-used symbols and letters which may be purchased for $5.

Diagrams may be drawn in black ink either directly on the model sheet, or on a separate sheet and pasted with rubber cement into spaces left for them in the text. Ballpoint pen is *not* acceptable.

Page Headings (Running Heads) should be centered, in CAPITAL LETTERS (preferably), at the top of the page – just above the blue line and touching it.

    LEFT-hand, EVEN-numbered pages should be headed with the AUTHOR'S NAME;

    RIGHT-hand, ODD-numbered pages should be headed with the TITLE of the paper (in shortened form if necessary).

    Exceptions: PAGE 1 and any other page that carries a display title require NO RUNNING HEADS.

Page Numbers should be at the top of the page, on the same line with the running heads.

    LEFT-hand, EVEN numbers – flush with left margin;

    RIGHT-hand, ODD numbers – flush with right margin.

    Exceptions: PAGE 1 and any other page that carries a display title should have page number, centered below the text, on blue line provided.

        FRONT MATTER PAGES should be numbered with Roman numerals (lower case), positioned below text in same manner as described above.

MEMOIRS FORMAT

> It is suggested that the material be arranged in pages as indicated below.
> Note: <u>Starred items (*) are requirements of publication.</u>

Front Matter (first pages in book, preceding main body of text).

    Page i – *Title, *Author's name.

    Page iii – Table of contents.

    Page iv – *Abstract (at least 1 sentence and at most 300 words).

        *1980 Mathematics Subject Classifications represent the primary and secondary subjects of the paper. For the classification scheme, see Annual Subject Indexes of MATHEMATICAL REVIEWS beginning in December 1978.

        Key words and phrases, if desired. (A list which covers the content of the paper adequately enough to be useful for an information retrieval system.)

    Page v, etc. – Preface, introduction, or any other matter not belonging in body of text.

Page 1 – Chapter Title (dropped 1 inch from top line, and centered).

        Beginning of Text.

        Footnotes: *Received by the editor date.

                Support information – grants, credits, etc.

Last Page (at bottom) – Author's affiliation.

ABCDEFGHIJ–AMS–898765432